T0250070

Constructing the countryside

Terry Marsden
University of Hull

Jonathan Murdoch
University of Newcastle upon Tyne

Philip Lowe
University of Newcastle upon Tyne

Richard Munton
University College London

Andrew Flynn
University of Hull

Routledge
Taylor & Francis Group

LONDON AND NEW YORK

© Terry Marsden, Jonathan Murdoch, Philip Lowe,
Richard Munton, Andrew Flynn 1993

This book is copyright under the Berne Convention.
No reproduction without permission.
All rights reserved.

First published in 1993 by UCL Press.
Second impression 1996.
Reprinted 2004
by Routledge,
2 Park Square, Milton Park, Abingdon, Oxon, OX14 4RN

Transferred to Digital Printing 2004

The name of University College London (UCL) is a registered
trade mark used by UCL Press with the consent of the owner.

ISBN:
1-85728-040-7 PB

A CIP catalogue record for this book
is available from the British Library.

Typeset in Palacio (Palatino).

CONTENTS

PREFACE

This book is the first in a series reporting on our work conducted under the auspices of the UK Economic and Social Research Council's Countryside Change Initiative (1988–93). The work assesses the processes of rural change in the advanced economies, with a focus on the British experience. It is aimed at a wide audience of researchers, graduates and undergraduates in a range of social science disciplines. This first book presents our conceptual approach to the study of rural change. Its aim is to redirect and re-invigorate social science enquiry into questions of rural development in advanced economies by developing a new perspective on the nature of change and the methodological tools necessary to investigate it.

Amongst the many colleagues who have given us their support we are particularly grateful to Jacquie Burgess, Fred Buttel, Graham Cox, David Goodman, Carolyn Harrison, Norman Long, Michael Redclift, Neil Ward and Sarah Whatmore. We are particularly indebted to Julie Grove-Hills, who worked with us for three years on the Cumbria case study under the Countryside Change Initiative. We would also like to thank Patsy Healey for allowing us to reproduce Figure 5.1. Thanks are also due to Mary Anne, Joseph and Hannah Marsden, Suki and Jake Flynn, and David Shields who typed every line (including this one) and helped administer the project.

TERRY MARSDEN   JONATHAN MURDOCH   PHILIP LOWE
RICHARD MUNTON   ANDREW FLYNN
September 1992

CHAPTER 1

Rural restructuring

Introduction

The postwar boom petered out in the 1970s. Its demise ushered in a period of upheaval worldwide, and we are still living with the consequent political and economic changes. Because these were largely unforeseen, they have presented a profound challenge to established ideas and theories about the nature of modern society. A major intellectual effort directed towards establishing a new understanding has followed. Agreement on which social processes should command our attention and how they should be treated is most unlikely. Even so, the research agenda will almost certainly include analyses of the increasing mobility of capital, the adoption of more flexible production methods, the complex relations between technology and environment, the influence of more clearly articulated consumer interests, and the widespread deregulation and reregulation of economic and political structures.

The combined effects of these global tendencies are major sources of uncertainty for nation states, local communities and families alike, undermining regional economic and political stability beyond the major trading blocs and fostering greater diversity of local experience within them. They are also eroding traditional disciplinary divisions within social science. If nothing else, the contemporary processes of global restructuring emphasize how the political, the social and the economic interact within and between different local, national and international spaces. No one process is dominant.

These tendencies challenge all aspects of social enquiry, not least that to which this book is directed – the rôle of rural space in the restructuring of advanced capitalist economies. For our enquiry, they pose two more specific questions. First, do they suggest a significant and growing rôle for rural space in the

1

emergent social and economic relations of modern societies? Secondly, if they do, what kind of conceptual and methodological framework would best serve such an enquiry?

We argue strongly that there is an urgent need to draw the study of rural areas and issues out of the margins and into the mainstream of social science, to reflect the contemporary economic and social salience of rural space. Crises of accumulation in capitalist societies necessitate the periodic and radical restructuring of production processes in order to establish new opportunities for profitable investment; one consequence is a reassessment of resources and spaces once considered unproductive or marginal. For a number of reasons, some rural areas once thought of as quintessential backwaters of economic activity have come to be seen as investment frontiers.

We will suggest, for example, that from the point of view of production, rural space is often attractive to capital, being less encumbered by earlier Fordist labour processes and rounds of investment; offers many new and more pleasant places in which to work and live than represented by the modern city and suburbia; and has become much more accessible as a result of improvements in telecommunications and transportation systems.

Rural areas have also long been repositories of small-firm entrepreneurship, which is now seen to be a key source of economic dynamism and innovation. At the same time, some of the new wave technologies, particularly biotechnology and information technology, are seen to favour rural locations. As Howard Newby has put it: "for the first time since the industrial revolution, technological change is allowing rural areas to compete on an equal basis with towns and cities for employment" (*Financial Times*, 8 December 1989).

In terms of consumption, especially among those large and influential sections of affluent societies that now place a high priority on non-material and positional goods as well as the accumulation of assets, rural space provides many sought-after opportunities, such as for living space, recreation, the enjoyment of amenity and wildlife, and a wholesome and pleasant environment. Beyond these are considerations of deep-rooted cultural and symbolic significance that two centuries of industrialization and urbanization have not diminished. As Williams (1973: 296) reminds us: "there is almost an inverse proportion, in the twentieth century, between the relative importance of the rural economy and the cultural importance of rural ideas". Our argument, however, is far more significant than simply to suggest the re-emergence of certain gemeinschaft notions concerning the "quest for community", or the historical maintenance of a rural idyll. It focuses instead on how, within an increasingly internationalized and service-oriented economy, the constant repositioning of rural issues, ideas and opportunities in the national polity and economy is a central feature of contemporary capitalist development.

The urgent need for a new conceptual framework arises not only because of the nature and significance of these changes, but also because rural issues have been historically marginalized in social science. Previous research methodologies are inadequate as starting points. On the one hand, the rural community studies of the 1950s and 1960s failed to engage with a wider world while, on the other, the Marxian political economy of the 1970s treated rural people and areas merely as passive recipients of the vagaries of national and international forces. Moreover, elsewhere in academia, government departments and non-governmental organizations (NGOs), the study of the rural economy has been almost exclusively restricted to inward-looking analyses of agriculture. Conducted through the medium of neoclassical economics, as if family farming businesses could be reduced to the rigorous assumptions of such an approach, and as if agriculture were all that the rural represented, these analyses have produced a partial outlook that is of diminishing relevance to advanced capitalist societies. Such an approach may now being broadened but only under the pressure of events, and especially the decline in the postwar food production imperative. It has therefore largely failed to relinquish its agricultural orientation, from which perspective other social and economic demands on rural space are perceived as being merely a function of agricultural weakness.

Only when the wider demands on rural areas are placed centre stage will engagement with the ideas of mainstream social science become the natural way forward. As we outline below, this way forward depends upon a revised understanding of the interaction between the social and the spatial as well as the derivation of new questions of rural change. This change of outlook means that rural sociologists, the disciplinary group most centrally concerned with these issues, must also go beyond traditional agrarian concerns, which focus upon the political and social position of agricultural labour. They must embrace the position and rôle of rural people, notions of rurality in contemporary society, and the processes and structures through which access to and use of rural resources are constructed. These are quintessentially *social science* questions.

It will also be argued that the nature of production and consumption in rural areas shapes and is shaped by both the labour process and the organization of property rights. New types of production are engendering new forms of labour relations (Marsden et al. 1992) and both rely upon the changing nature of property rights. These structuring mechanisms are critical to the reproduction of power in rural areas. What is crucial to rural economic restructuring in particular is the conjuncture between external, mobile capital and the distribution of local property rights. Productive capital still needs access to property rights, and at as low a cost as possible; it also seeks partial and fleeting fixity so as to permit maximum flexibility of use. In the rural consumption sphere, however, many of the holders of property rights wish to regulate, through the local power structure, the location, quantity and quality of capital to be fixed.

The book thus focuses on a set of key themes concerning the changing position of rural areas and "rurality", and seeks to ground its methodological concerns within a case study of the UK in the late 20th century. These issues may be summarized in three broad questions:

○ How are international processes of economic and social restructuring being expressed and mediated within one nation state?

○ How is the state "regulating" rural change and to what extent does the late 20th century represent a break with the past?

○ How can conceptual advances in mainstream social theory be applied to the rural arena and, conversely, how can locally based social action be effectively incorporated into our understanding of uneven development?

New questions of rural change

In recent years, a small body of literature has sought to examine change in rural areas within advanced capitalist economies through a focus on economic restructuring. It has largely concentrated upon the importance of labour adjustment and the diminishing significance of agriculture and its associated property relations in conditioning rural social change (Rees 1984, Barlow

1986). Specifically, it has sought to challenge the seminal work of Newby, which is based on a Weberian analysis of social interests in rural East Anglia (Newby 1977, Newby et al. 1978). Instead, emphasis has been placed upon the nature of industrial (mainly manufacturing) firms seeking out new rural locations, often to take advantage of pools of relatively cheap labour, but more generally to lower costs of production in greenfield locations. The growing internationalization of capital has brought a greater degree of locational "flexibility", with the quality and cost of labour power assuming particular importance. As Urry (1984: 55) suggests: "as long as there could be sufficient labour in a 'rural' area then expansion may well take place in that [greenfield] site rather than in alternative areas. Cities have become relatively less distinctive entities, bypassed by various circuits of capital and labour power".

It is therefore not surprising that in the USA, and the UK and other parts of western Europe new firm formation rates were higher in small towns and rural areas than in large urban centres (Keeble et al. 1983, Fothergill & Gudgin 1982, Hodge & Monk 1987, Champion & Townsend 1990). Increasingly, in more urbanized regions, service activities have also relocated in rural areas, thereby accentuating an employment pattern already heavily weighted towards the service sector. Another postwar feature has been the relocation of large manufacturing plants, or the expansion of public sector activities requiring remoteness (e.g. the defence industry and nuclear power stations).

More recently, some rural areas have been seen as contexts ideally suited to flexible, accumulative strategies for small businesses in a post-agricultural, post-industrial world. As Paloscia (1991) points out with reference to Tuscany, small businesses, both related and unrelated to agriculture, and in particular the existence of conducive social and cultural preconditions can provide the stimulus for flexible industrial development. Indeed, within many of the core areas of northern Europe and North America, rural spaces are now considered to provide amenable social conditions (i.e. a predominance of small-scale enterprises, family businesses and a cheap and adaptable labour supply) within which a diffuse system of production and service provision can flourish.

These transitions may be highly contested and locally variable. They are frequently being moulded by local actors and regulatory authorities, the outcomes of their interventions challenging facile generalizations about the attractiveness of rural locations for industrial activity. In the UK, for example, the growth of a residential middle class in country towns has been primarily responsible for drawing into them employment in personal and commercial services and public administration, adding to the traditional employment of tourism, retailing and the rural professions, while that middle class has resisted where it can the intrusion of "urban" forms of manufacturing activity. Instead, the availability of relatively cheap female labour has also encouraged the decentralization of administrative and clerical work. Moreover, because of their accessibility, environmental attractions and availability of highly skilled manpower, favoured rural regions (especially East Anglia, North Wales, the South and the South West) have drawn in employment in scientific, technical and financial services.

The empirical work so far conducted has frequently taken a partial view of the restructuring process, limiting its attention to economic and often only employment trends derived from aggregated statistical sources. This approach is unsatisfactory since it tends to subordinate rural economic and social change to the broader capital and employment restructuring processes previously described. It assumes a "top-down" causal argument when seeking to explain the uneven pattern of development, reducing rural areas to uniform and passive spaces upon which past, present and future rounds of capital investment engender radically different spatial divisions of labour. From this perspective, even if by default, rurality is largely seen as a descriptive and marginal category lacking explanatory power, and in which the variability and significance of local social action is ignored (Chs 2 & 6).

The emphasis placed on "urban" and "regional" frameworks of analysis has obscured a thorough understanding of restructuring in the rural context, beyond a tacit recognition that a trend towards flexible systems of production allows firms greater opportunities to establish and rearrange their organization beyond their previous spatial confines. We wish to suggest, however, that there are common sets of particular structuring mechanisms that operate in rural areas. These extend beyond notions reducible to economic terms and take the debate into social, political and ideological spheres. They also encompass a range of factors associated with relative access to property, as well as labour-market participation and opportunities, and include the historical particularity of state action in rural areas. These mechanisms, and appropriate means of investigating

6

them, have to be integrated into a broader notion of restructuring if our understanding of the transformation of rural areas is to be rescued from a conceptual hiatus. It is imperative to respond to Newby's (1986) call for a holistic analysis of rural social relations. In the first instance, it is necessary to explore the common conditions of contemporary rural development in advanced economies. To varying degrees, rural areas share a legacy that springs from the social and economic relations of agriculture and other forms of primary production. This is the basis of the distinctiveness of their labour markets, which are characterized by local conjunctions of working-class quiescence, petit-bourgeois ownership of capital and small-scale enterprises (Bradley 1985). The postwar technological revolution in farming has transformed its labour process and diminished its overall significance as a major employer of rural labour. Nevertheless, as Whitener (1989) argues for the USA, and Newby's studies demonstrate for eastern England, agriculture's historical predominance still has important implications for the development of rural areas, conditioning the comparative advantages or disadvantages they offer to other fractions of capital. A predominantly agricultural local labour market can be seen as advantageous for new firms wishing to relocate in relatively low-wage and non-unionized areas. Agricultural labour shed today from small, industrialized and technologically sophisticated farm businesses may also offer a wide range of transferable, practical skills.

As well as the labour opportunities provided by rural areas, their structures of simple commodity production can provide the conditions for rapid economic adjustment because they are characterized by small-scale, flexible accumulation and specialization, and a culture of entrepreneurship. State support for agriculture, moreover, has been capitalized in land values, giving rural landowners significant sources of collateral when they seek to develop and expand new businesses. Thus, in some rural areas, particularly those that form part of economically buoyant regions (such as East Anglia, Bavaria, Tuscany, Colorado and New England), the spatial and social structures established around agriculture and other forms of land-based production can offer advantages to both producers and consumers in the shift towards more flexible systems of production and service provision. Not only have such areas come to be seen as having a more attractive physical environment than the old industrial towns and cities, but also a more amenable socioeconomic environment, composed of small family businesses and a placid labour force little marred by sharp social conflicts or the so-called dependency culture.

Such shifts, however, may not be uniform or sustainable. To the extent that they have depended on high levels of public expenditure in agriculture, infrastructure, regional development, public services and defence, they are vulnerable to the cuts in public spending precipitated by neoliberal economic policies and the ending of the Cold War. At the same time, the increasing integration of rural areas into the world economy has increased their exposure to the vagaries of international markets, business cycles, shifts in production technology and, eventually, work practices. After all, rural areas in the advanced economies are but one type of locational option in the global space economy, and their wage rates, though comparatively low by the standards of advanced economies, are rarely competitive with those in the Third World (Dicken 1992).

Much depends on the interaction between the economic and political contexts. As Lawrence (1990) demonstrates for Australia and Summers et al. (1990) and Falk & Lyson (1989) for the USA, an over-reliance upon agriculture, combined with neoliberal macroeconomic policies directed towards deregulation and the national removal of tariff barriers, can expose rural areas to the full force of international competition in commodity markets. In both cases, national policies designed to accelerate productivity-oriented technological development in agriculture are further reducing the demand for farm labour. They are also redefining work rôles, local labour conditions and working practices, often with devastating social consequences, while reductions in social welfare provision have given rise to a rural "underclass".

Thus, although agriculture may no longer dominate the economy in many rural areas, the degree of historical and contemporary reliance upon it, often artificially sustained by state support, is still a principal conditioning factor for new rounds of investment. Even if farming represents only a residual element in the rural economy, it often retains a disproportionate social and ideological significance in the moulding of social and economic change through the politically entrenched positions held by farmers and landowners. Their power as a political fraction may be locally variable and in the long term subject to historical decline, but through their involvement in village, county and national politics it can extend far beyond their local control over land.

Furthermore, access to property rights, and not necessarily all the bundle of rights attributable under common law to freehold ownership (Ch. 4), remains a major source of power and prestige in rural societies. The structure of ownership can significantly affect the ways in which new capital is fixed to property, allowing the previously dominant fractions of landed capital to continue to hold much greater sway over rural change than their contemporary national significance might otherwise suggest. In particular, although the current productive function of agriculture may be set to retreat, established farm families and other landowners will continue to act as significant gatekeepers in affecting the timing and pattern of access to rights by others. Rights to housing, and to amenity and industrial development, will all be partly mediated by the owners and occupiers of agricultural resources, even if they now constitute a tiny proportion of the rural population. Moreover, access to land and property tends to reinforce patterns of deprivation and wellbeing established in the labour market.

As the primacy of agricultural production (as *food* production) diminishes, new consumption-oriented rôles, such as recreation, leisure and environmental conservation, as well as other primary-production activities such as biomass and mineral extraction, are growing in prominence. In turn, these changes create new and locally specific accumulation opportunities, making rural areas the locus for new social and political conflicts. As Mormont (1990) argues, with increasing levels of mobility and new uses of the countryside, particularly by urban and ex-urban residents, the significance of the "rural" has come to be associated less with belonging to a particular place (Cohen 1988) and more with the varying levels of opportunity that rural areas afford. He draws attention to the powerful symbolic and ideological significance contained within the notion of "rural": it is most effectively understood today as an active set of "representations" based upon competing and often conflicting principles linked to certain styles of living, working and recreation. Central to his position is the fact that spatial and cultural changes have enhanced the significance of consumption as a source of identity and political conflict. Consumption of the natural environment has become a distinctive feature of spatially extensive life styles associated with certain mobile and influential social strata. As we shall see in later chapters, the rural land development process (i.e. the methods and social actions surrounding the exploitation of land for different functions) forms a key arena in which these representations are mobilized and through which access is regulated to increasingly scarce resources.

The growing recognition that different forms of rural experience (in terms of leisure, recreation, housing, employment and heritage seeking) are desired attributes of an "urban" existence and of culture means that rural areas now attract different exploitative interests aiming to serve these growing markets. For example, the corporate house-building industry seeks out available land for "exclusive" housing, and the leisure industry purchases land for golf and other sporting complexes. Such interests are guided in their investment decisions both by the increasing demand for "rural" pursuits, experiences and values, and by the historical attractiveness and "authenticity" of particular rural places. They transform rural areas as they simultaneously seek to reproduce their scarcity, heritage, uniqueness and desirability as refuges from a chaotic, urban world. Some areas are left aside, while others are transformed in appearance, labour market conditions and social structure. The nature of local endowments may have changed, but their importance remains.

The most significant economic interests investing in rural areas, as elsewhere, are associated with corporate organizations at national and international scales, including agribusinesses, house-builders, mineral and industrial firms, forestry companies and leisure concerns. Their interest in rural society is restricted and their long-term commitment to siting plant and investment in specific areas cannot be guaranteed, since they are preoccupied with the endless quest for advantageous locations. Their strategies are often linked to internal and transnational reorganization, asset realization and high levels of credit leverage (Clark 1990), and their investment patterns are determined (as in the rural housing sector) by speculative development during upturns in the business cycle. The financial deregulation and economic uncertainty of the past decade merely presage even greater fluctuations in the demand and supply of credit, increasing the volatility in amount and quality of capital invested in particular rural areas. Thus, although it may be argued that small firms of local builders, developers and businessmen operating in the interstices of the market can facilitate endogenous rural development in *some* places, it is the strategic interests based nationally and internationally that set the pace and direction of change. Dependent themselves on merchant credit, these enterprises are increasingly geared to opening up market opportunities in what have traditionally been areas beyond their reach; it follows that local actors have to be increasingly adept at responding to the speed and multidimensionality of the changes that originate in a larger and dynamic world.

For example, more deregulated and credit-dependent "boom and bust" cycles in agriculture and housing development have occurred in recent years, with attendant high levels of risk and uncertainty in fixing capital to productive uses. Moreover, although traditional forms of extensive production (agriculture, forestry, mining) have tended to suppress short-term speculation and the effects of market and credit volatilities, partly because of their relatively long investment and production times and because of state support for their respective commodity markets, the newer forms of capital investment, such as that in industrial plant, housing and leisure development, enjoy no such in-built constraints. They are much more strategically tied to a "just-in-time" level of market sensitivity, with the length of their production periods reduced through the application of novel building technologies, flexible and peripatetic working practices, and new techniques of project management. Coping strategies designed to mediate the higher level of uncertainty may introduce new forms of property rights that minimize the fixity of capital and reduce the level of risk (Ch. 4).

The relationships between capital and property rights are complex. Now, as in the past, land and the rights associated with it perform a variety of functions in the different production processes. The "natural" properties of land and space are converted into exchange values and commodities in an uneven way according to the type of exploitative activity. For instance, the peculiar position of land in agricultural production (as both a condition and a means of production), as well as its restricted supply, has perpetuated monopoly farming and landowning interests in land rights. In industrial and housing development, land forms not only the site for production but also the potential for varying levels of valorization as development gains are made with the consent of local planning authorities.

Moreover, a wide range of property rights can be developed during the land-development process, including initial options and conditional contracts established between landowners and developers, as well as forms of leasing and joint-ownership arrangements. Such rights largely codify the extent to which the different interests acquire opportunities to exploit production processes at different times. For example, mineral-extracting companies seek mutually beneficial property arrangements with landowners, whereby ownership, occupancy and user rights are developed on a flexible basis, in response partly to the incidence of planning permission for extraction and partly to fluctuations in demand. The parties concerned create new rights through the rearrangement between them of existing rights over variable time periods. These agreements enable them to share the costs and maximize the benefits of their different exploitative priorities. Insofar as it is possible, they will also ensure a restricted distribution of the development gain, leaving the non-landowning rural residents excluded from it, even if they experience costs arising from the development and subsequent use of the land. It is at this point that those who do not hold a direct, private interest in the property concerned will seek to impose their indirect interest via the public regulatory system; it is part of our argument that these indirect interests, often associated with rural consumption, are increasingly able to affect the local distribution of property rights.

The state, if not the civil law, is the ultimate mediator of the contestation over the redistribution of property rights, and it is the state that must also traverse the highly contested shifts in social meanings and representations that rural restructuring entails. These requirements have involved it in contradictory relationships with property rights: on the one hand protecting and promoting them as expressions of fixed capital, on the other challenging and overruling them where they obstruct capital mobility and new rounds of investment.

At the same time, the increasing mobility of capital and labour, facilitated by technological and infrastructural developments and the integration of national and regional economies, has undermined the traditional basis for the spatial management of economic and demographic change. New approaches to economic policy adopted in pursuit of economic restructuring have, in turn, involved some territorial restructuring of states, partly to facilitate more flexible accumulation. Pickvance & Preteceille (1991: 214) identify the following common trends in advanced capitalist societies since the mid-1970s:

> the increasing level of local government mobilization in support of (private sector) economic development; the decreasing rôle of central grants in financing local government spending; the growth of neo-liberal tendencies in public policies leading for example to the privatisation of collective consumption services and to the spread of market relations as a mode of regulation; the parallel weakening of welfare state values such as equality, social justice and

redistribution; and an increase in territorial social inequalities and strengthening of social segregation in space at the regional level as well as between and within towns.

With the return of mass unemployment since the mid-1970s, central governments have sought to disavow responsibility for macroeconomic management. According to the new orthodoxy, a country's welfare has depended on the position of its industries in the world economy and on the wage levels paid in each industry. Unemployment has been attributed to wage inflexibility, and should, first and foremost, be the concern of individuals and producer groups, not of government. The rôle cast for government is that of removing supply-side constraints and promoting an enterprise culture, encompassing a rhetoric that talks of getting the government "off the back" of business, rolling back the state, and freeing-up markets. However, in order to effect such profound changes, the state has had to take on an active, interventionist rôle, including the deployment of public resources to support private capital accumulation, the promotion of private enterprise in the provision of collective consumption, and the displacement of values such as social justice and equity by a belief in the economic and social efficiency of the market. The involvement of national governments in economic restructuring is therefore ambivalent. Partly in response to this, there has been a marked tendency internationally for local government to assume an expanded rôle in local economic regeneration, as well as the development of transnational structures.

Within this context, the territorial management of rural areas has faced a number of specific challenges. Traditionally, rural policy has been dominated by agricultural productivism and its supporting planning policies, with the institutions of agrarian corporatism instrumental in rural development. For more than a decade now, productivism has been in crisis, beset by many contradictions that have fuelled pressures for agricultural policy reform. Coincidentally, support policies of key settlement planning and regional planning have been undermined by counter-urbanization and economic restructuring.

The increasing insinuation of non-farming interests into rural areas and the build-up of pressures to liberalize farm supports signal the demise of agrarian corporatism as the key instrument of rural management. In its stead there has emerged a fragmentation of localistic orientations as individual rural communities and areas express their specific consumption or rural development needs. The consequence is heightened differentiation of rural areas, which governments themselves have wittingly or unwittingly encouraged. On the one hand, anxious to curb agricultural supports and surpluses, they have devised various programmes to encourage farmers to diversify their businesses or to leave the industry. On the other hand, there has been a general abandonment of strong regional development policies. The curtailment of gross depopulation from many rural regions has meant, moreover, that new justifications have had to be sought for rural development programmes, based on arguments to do with the inequalities and opportunities rural areas present, rather than their marginality or backwardness. There has been a consequent shift in rural regional policies from state-led industrialization programmes and inducements for inward investment to the promotion of indigenous small firms and self-help initiatives based on endogenous development models (Pettigrew 1987, Day et al. 1989, Cloke 1990, European Commission 1988, O'Cinnéide & Cuddy 1992).

Thus places increasingly compete in attracting private and public investment, striving to present their specific mix of human, cultural, financial, physical and natural capital in a favourable light, with a distinct identity. Equally, they seek to defend these attributes from unwanted development, although what counts locally as wanted or unwanted development has become a matter of growing contention. Social and cultural tensions arise from the different visions and expectations people have about the same place, reflecting their separate life styles and livelihoods. The tensions over the relative needs of an area for economic development and environment protection are among the most acute.

This increasingly localistic focus in rural management has been paralleled by the development of transnational structures of governance. In Europe, for example, the processes of economic restructuring are being accompanied by probably influential major institutional changes associated with the unification and extension of the European Community (EC). The move towards a Single European Market will lead to further integration in production, distribution and consumption networks. Regulation by the EC in such diverse fields as competition policy, working conditions, environmental standards and agricultural support has introduced a new tier of rule formation and compliance by local and national institutions. A parallel challenge to national sovereignty is posed by the EC effort to strengthen the European Parliament, accelerate moves towards a federal structure, and support greater institutional and political autonomy at the regional level. Regional integration is also well illustrated by the 1992 agreement between the USA, Canada and Mexico to form the North American Free Trade Area, a conglomerate with 360 million consumers that is large enough to compete on equal terms with the EC and a South-East Asian economic grouping based around the Japanese economy.

These trading blocs provide an alternative, if not necessarily complementary, approach to the current search for global agreements to stabilize the world economy and manage the Earth's resources. The main initiatives include: the current GATT talks, where agricultural policy reform is seen as pivotal, but which include important negotiations concerning such areas as intellectual property rights and financial services; the 1992 Rio de Janeiro Conference on global environmental problems; and the G7 or IMF meetings to secure international financial stability. This new combination of regional, European and global governance poses an entirely new challenge to conventional assumptions about the nature and geographical scale of public intervention and regulation of the economy.

There are indications, however, that the whole project of global integration has reached its limits, as indicated, for example, by the failure of the Rio Conference to bring a sense of urgency to the collective management of the planet's environment, by the impasse in the Uruguay Round of GATT negotiations on trade liberalization, by the difficulties encountered in ratifying the Maastricht treaty on EC economic and political integration, and by the gathering opposition to fixed exchange rates and transnational co-ordination of monetary policy as the world recession continues.

This is not to suggest that the internationalization of the global economy will not continue, nor that attempts to regulate it will not occur, but international capital may find itself more constrained by national self-interest and local objections than international regulatory action. Local objection will be especially evident in densely populated countries such as the UK, where local property rights have been long established. Alongside them are important regulatory bodies, such as the local planning authorities, whose land-use decisions are increasingly significant in creating differentiated rural spaces. This is because many of the new rural interests identified in this discussion remain, to varying degrees, land-based, and this makes their development decisions directly subject to local action through the planning process. Moreover, these decisions will be multidimensional, reflecting the wide-ranging competition for rural resources from, *inter alia*, housing, amenity, conservation and economic development interests. They will no longer be agriculturally led, except perhaps in those rural spaces specifically identified for further expansion of industrialized farm production; nor will they be exempted from planning control in the way that agriculture in the UK has been for the latter half of this century. The sheer range of land-based interests involved ensures that rural development will become a central focus and dynamic in national culture and polity.

This volume is concerned with the various dimensions of rural restructuring. These are variable in both time and space. We have therefore chosen to ground much of our discussion by reference to the rural UK (Chs 3, 4, 5). This empirical focus allows us to examine what the key aspects of restructuring will be (Ch. 2) and to consider how such space is shaped by them. Chapter 1 has provided an introduction to the main research questions:

○ How are the international processes of restructuring expressed within the nation state and how does the state attempt to regulate these processes?

○ What is the relationship between the economic and political contexts and how is the restructuring process helped or hindered by state policy?

○ How do broad restructuring processes affect different spaces (for instance, localities), what regulatory tools are available within such spaces to constrain specific accumulation strategies geared towards opening up market opportunities, and does this lead to a fragmentation of localistic orientations as individual places "express" their consumption or development needs?

One theme running through what follows is "differentiation". It is our view that rural places are becoming more differentiated from one another as the complex processes of restructuring weave through both time and space. This forces us to take seriously how we conceptualize this new multiplicity of rural formations (Ch. 6) and what the most appropriate methods for their analysis might be (Ch. 7). We are concerned here to provide more than simply the historical background to the current situation; we wish also to present a set of analytical tools that will allow us to make sense of it.

CHAPTER 2

Restructuring the countryside
Key conceptual developments in assessing rural change

Introduction

There can be little doubt that "restructuring" has become one of the commonest and most overused terms in social science. It has been widely employed to signal a distinctive break in the progress of many capitalist economies in the 1970s and 1980s, and the ensuing social and political consequences. At the global level, these changes are usually associated with the oil shocks of 1973 and 1978, and the demise of US economic hegemony, which is associated with the inability of the dollar to sustain fixed exchange rates as agreed at the Bretton Woods Conference in 1944. The postwar boom was palpably over: many national economies were beset by high inflation and unemployment rates not seen since the 1930s. The overall rate of growth in the world economy declined substantially during the 1970s, and, just as important, its international pattern was altered. The concomitants were profound: they included the relative decline of manufacturing industry in advanced economies; the emergence of new technologies, ranging from information technology (IT) to biotechnology; the increased globalization of capital movements; and the acquisition of power by neoliberal politicians committed to overturning Keynesian orthodoxies in favour of fiscal rectitude, low inflation and free markets.

By the early 1980s these changes were awakening social scientists to the poverty of their knowledge about the nature and sustainability of contemporary economic development. The conceptual tools at their disposal – developed under presumed conditions of social stasis, political consensus and the continuity of economic growth – suddenly seemed outmoded; and this realization fuelled a critical perspective on social change. Progressively, different groups of scholars in the Western world began to reject more firmly than before traditional disciplinary boundaries, attempting in particular to apply a neo-Marxist critique (often informed by Third World development studies) to the rapidly evolving situation. The pace and depth of change, though constantly threatening to overwhelm attempts to rethink social theories, has nonetheless stimulated a wide-ranging debate on the causes and consequences of restructuring.

Three principal avenues of inquiry have emerged. First, and broadest, has been the "regulationist" approach initiated by a group of French scholars but subsequently adopted and elaborated upon by many throughout Europe and North America. It addresses the central question of what were the macroeconomic and institutional underpinnings of the unprecedented rate of economic growth experienced by most advanced industrial economies between 1945 and 1970, and what brought this growth to an impasse (Aglietta 1979, Lipietz 1987). Growth was seen to have been based on the international diffusion of an American, Fordist model of industrial development incorporating appropriate techniques and labour processes for mass production. The successful and systematic deployment of this model, it is argued, depended on the mass consumption that went with it, realized through the growth of disposable incomes and increasingly global markets. In addition, a critical set of norms and policies that maintained a congruent relationship between the growth of production and of consumption helped to sustain and legitimate the high level of economic growth. These consisted of such monopolistic forms of regulation as Keynesian economic management, the welfare state and collective bargaining.

From this perspective, the present is diagnosed as a period of significant change and crisis. One long cycle of growth (or regime of accumulation) and its institutional and social support systems (mode of regulation) have waned. Although new structural tendencies (see below) have emerged, the regulationist school considers that these have not become sufficiently consolidated to establish clearly the character of the successor period. A second school of theorists is less equivocal and more controversial, regarding the present as a major historical divide that marks the emergence of a new era based around "flexible specialization" in industrial production. The new era depends upon the increasing availability of generic and highly flexible technologies, and the proliferation of market fragmentation and customized demand. It is variously characterized as an era of "disorganized capitalism" or "post-Fordism" (Piore & Sabel 1984, Lash & Urry 1987, Gertler 1988, Gertler & Schoenberger 1992, Hirst & Zeitlin 1991). Its governing principles are seen to be non-standardized demand for goods and services, and vertically disintegrated production, based upon the decentralization of tasks and responsibilities to loosely interconnected units. Inevitably, the centralized, monopolistic structures and standardized, universal forms that characterized mass production and mass consumption are regarded as increasingly obsolete features of a past system. The spatial and sectoral unevenness of the substitution of the new for the old shows how ambiguous such terms as "post-Fordism" and "post-modernism" are, and points up the weakness of unilinear arguments over economic development such as those underpinning, for example, Marxist political economy.

A third approach to restructuring is associated with arguments surrounding the notion of "spatial divisions of labour". The thesis was initially put forward by Massey to provide a new conceptual framework that would account for the changing geography of capitalist accumulation. It focused in particular on the changing rôles and relations of regional and local economies in the processes of accumulation (Massey 1978, Massey & Allen 1988, Walker & Storper 1986). More than other approaches, it places emphasis on the greater geographical variability in new investment implicit in the decline of Fordist industrial organization, and the ascendancy of more flexible systems of production and exchange. From this perspective, individual localities are seen to be the result of the interaction between the legacy of their former economic relations and their contemporary economic rôles (Ch. 6). The work of Massey, and Lash and Urry in particular, stresses spatial differences in the organization of production, but having emerged from a primary concern with regional industrial development they rarely give attention to rural areas (but see Bradley 1985, Summers et al. 1990).

Quite separately, from the late 1970s, the political economy of agrarian development in advanced societies became an active field of study. Its proponents also sought to provide a more holistic analysis of the international food system, of the rôle of the state in mediating capitalist agrarian development, and of the position of "non-capitalist" forms of agrarian production within monopoly capitalist societies (Marsden et al. 1986, 1992, Buttel et al. 1990). This was part of a deliberate move away from a territorial perspective, but there was little interaction with the restructuring approaches despite initiatives to place rural change within broader trajectories of economic development (Bradley & Lowe 1984). Whereas restructuring theorists have tended to marginalize the rural, as only one territorial space upon which the transformatory surges of capitalist reorganization were to be played out, the agrarian political economists prioritized the systematic study of the food chain in which the decreasing relevance of land-based production was a central theme (e.g. Goodman et al. 1987). Little attention has thus been paid to the integration of these aspects of rural change. The lack of analysis of the changing position of rural areas, and people living and working within them, has become more and more apparent with the recognition that rural change is deeply embedded within restructuring processes more generally. In a variety of ways, rural areas provide new opportunities for restructuring and there is therefore a clear need to redress this neglect.

In seeking to take the debate forward we are especially mindful of the need to move beyond the broad theoretical concepts that dominate much of the restructuring literature. It is equally essential to develop the means to conduct programmes of empirical work that focus upon explaining processes of change as they are experienced at the local level. These advances will, of necessity, question some of the existing theoretical concerns. For example, current notions within the literature, emerging as they have very largely from a political economy perspective, tend to retain an excessive economism and a set of "top-down", structuralist assumptions about the nature of change. They make insufficient allowance for either local action or non-material considerations in explaining the movement, fixing and accumulation of capital. Furthermore, little attention has been paid to the rôle and status of either the nation state or the local state within an internationalized set of economic and political circumstances (McMichael 1992). In the rural sphere, in particular, debates on the persistence or otherwise of the peasant farmer as a petty commodity producer, have sharpened criticism of the inadequacies surrounding structuralist analyses of social action (Smith 1986, Scott 1986). Moreover, the need to appreciate the significance of space and locality not just as residual variables but as causal social factors in moulding development has brought more urgency to the attempts to link together structures and local action (Ch. 6).

In an attempt to bridge the gap between theory and empirical enquiry, and to acknowledge our concern for local social action, we analyze four middle-level analytical concepts. Taken together, we would argue, these constitute the most appropriate frame of reference from which to assess processes of social and economic change. The preferred approach to the conduct of empirical inquiry in a local setting, and its methodological basis, are discussed in Chapters 6 and 7. In this chapter we address the other four foci of our conceptual analysis. These are, first, the changing relationships between *production* and *consumption*; secondly, the *commoditization* of social and economic processes; thirdly, *representation* as a social and political process which continually redefines the arena in which contestation over resources occurs; and fourthly, the integration of *property relations* as a key structuring mechanism guiding change at the local level. The illustrations we select bear most obviously upon rural space but we see no reason why these concepts could not be applied equally to other kinds of space. Finally, at this early stage in our thinking, the precise relations between these concepts, and how they variably assist the explanation of change in different spaces, cannot be anticipated outside the historical and geographical specificities of the particular places in which empirical analysis is to be conducted.

Production and consumption

As we argued in Chapter 1, rural areas are sites of consumption as well as of production, and yet the restructuring literature concentrates on the relations of production, often exclusively on those of manufacturing. In agrarian political economy, likewise, emphasis has been on an understanding of the dynamics of production processes, with limited attention paid to social rigidities or changes in consumption practices. Most studies of housing (Merrett 1979, Ball 1983) are also largely production oriented, despite the increasing realization that deregulated private-sector activities are increasingly consumption led (Cooke 1989a, Saunders 1990, Barlow & King 1992). In the context of rural areas, it is particularly pertinent to consider the interrelations between production and consumption, given the increasing rôle of such areas as consumption spaces. The regulationist approach has been most advanced in this regard, arguing that the interdependencies of production and consumption are crucially mediated through the state. The rôle of the state will differ according to the nature of the prevailing regime of accumulation, which Lipietz (1985: 13) defines as "a systematic mode of dividing and reallocating the social product, which achieves over a period a certain match between the transformation of the conditions of production and final consumption".

Such an outcome needs to be actively managed in what Lipietz terms the mode of regulation. Modes of regulation, above all, are means of institutionalizing struggles between competing interests, containing them within bounds that reproduce and legitimate the desired balance between production and consumption within a particular regime of accumulation (Lipietz 1987, Jessop 1990). The state is thus ascribed a dialectical relationship with the process of accumulation.

For many regulationists, the decline of Keynesianism and the emergence, however unevenly, of a new mode of regulation associated with deregulation and privatization needs to be linked directly to the move away from the Fordist regime of accumulation (Thompson 1989, Saunders 1990). Since the mid-1970s, advanced capitalist economies have become less stable. Physical and human limits to production systems, and the inability to install technologies quickly enough to replace costly labour created a structural crisis that brought the long postwar cycle of growth to an end (Glyn 1989). The development of new products, markets and production "spaces" had become essential and necessitated new forms of social capital and infrastructure, and urgent modifications to labour conditions, if productivity and profitability were to be reinvigorated. At a political level, the response was advocacy of a free market ideology of "deregulation" and the extension of markets, combined with state intervention to reduce the corporatist power of industrial labour (Gordon 1980, Boccara 1985). Sustaining the new regime of accumulation, however, depended equally on the introduction of new modes of consumption. In the current case, these were to be based on niche markets and individualized consumption, whether associated with fashion, housing, entertainment, food or type of insurance arrangements. All were linked in some way to flexible specialization, and product differentiation (Hirst & Zietlin 1991). Thus, far from simply adding on a concern for consumption issues, it is necessary to see how these new modes reinforce our understanding of production and how the state provides the conditions for the maintenance of these new accumulation strategies (Featherstone 1990).

Friedmann (1988) provides an illustration of these arguments for the food sector. She attempts to relate historically contingent regimes of accumulation associated with particular food complexes (such as grain-fed livestock and meat production) to specific Fordist norms of consumption dependent upon class-based consumption habits (Ch. 3). Her case study reveals that although capital in the meat industry may have become more globally organized during the 20th century, its organization still remains dependent upon national and regionally constructed consumption patterns, indicating the crucial importance of the social regulation of consumption to the sustainability of any particular regime of accumulation.

A further example is provided by recent changes in the UK housing sector. The extension of market-oriented policies during the 1980s not only altered the nature of housing production and the opportunities for corporate capital in the housing development sector, but also redistributed the benefits from consumption. For example, the policies rearranged the ownership and legal control of both the means of production (through, for example, denationalization and deregulation) and of consumption (through opening up rights to purchase public housing at less than market prices). Moreover, Saunders & Harris (1990: 73) suggests that it is in the reorientation of the rights of consumers that these policies have achieved their full political and sociological significance. They conclude that:

Arguments for and against denationalization and liberalization [in the housing market] are important in economic terms but arguably have little consequence for a sociological analysis of social change. Commoditization and the creation of social market arrangements are where we should look for evidence of fundamental changes in the organization and experience of everyday life.

Of equal significance to the development of consumer-oriented policies through the deliberate extension of commodity markets has been the state's encouragement of, and a parallel technological revolution in, financial markets. These have led to the deregulation of credit provision, encouraging a rapid growth in personal debt, and a rise in the use of securities funds to stimulate industrial takeovers. The corporate house-building and food sectors, both with substantial interests in rural areas, have played a full part in these developments (Marsden & Whatmore 1992).

These economic and political changes will engender new rounds of investment which, in turn, will create new spatial patterns of production and consumption. But space cannot be treated as uniform; neither can *local* modes of regulation be assumed to be standardized or disinterested. In particular, the quality of local space, as well as its supply, can be regulated to the advantage of both production and consumption interests.

23

Specifically, the particular social and economic configurations of consumption in rural localities provide important signposts for capital investment. The relative "attractiveness" or "authenticity" (Urry 1990) of certain rural localities is associated with their consumption potential as well as their utility for production purposes. For instance, the specifically rural nature of executive housing demand is in itself a particular norm of consumption related to, and partly generated by, the corporate house-building industry. Local planning, by means of its control over land availability, fulfils a mediating rôle between the accumulation strategies of the house-building industry and the consumption norms of the new rural middle class, protecting the scarcity value, authenticity and positional status of exclusive housing (Thrift 1989, Elson 1986, Short et al. 1986).

Such norms of consumption are subject to constraint, which in turn, regenerates the norms themselves; people's consumption of scarce and finite positional goods is at once both highly visible and socially selective. Its very selectivity provides a driving force for the development industry. As Hirsch (1978) illustrates with reference to the suburbanization process, far from representing a voluntary set of market exchanges, these consumption practices (particularly when associated with scarce goods) can take on a coercive character for those involved. Recreating scarcity value and the authenticity of rural living depends upon the social maintenance of exclusivity. Keeping people out, and the political, economic and social processes designed to do this, thus provides the social conditions for others to consume. Aspirations to obtain scarce positional goods in the countryside collectively provide consumption norms which, in turn, attract exploitative development interests. Crucially, however, the satisfaction derived by "the consumer(s)" may not be infinitely expanded; it depends on the relative position of others. Scarce goods have to be continually protected and reconstructed through social and political processes. For example, at the local level, once an executive house is purchased, the "consumer" is obliged to pursue social and political means to protect the social value of the purchase.

The synergistic relations between production and consumption interests contained in this example of executive housing also reveal the conflictual nature of much rural development. Through the spatial structuring of development, for example, a benefit secured in one place can inflict a cost elsewhere. One illustration is provided by the growth of home ownership and thus the need for privatized (exclusive) space on the one hand but access to different types of public space on the other. Restrictive planning policies may encourage high densities of urban development, leading to a lack of urban open space, while the social benefits arising from these conditions may be bestowed selectively on residents of private rural housing whose lower densities and closeness to open countryside are thereby preserved (Marsden et al 1992). But, in turn, the latter's enjoyment of the countryside may depend partly on the assertion of collective amenity rights with regard to the aesthetic quality of the landscape and access to it, against the private rights of rural landowners to engage in intensive agricultural production. In this way, much rural development involves a succession of conflictual relationships between different places and between different spheres of production and consumption.

Recent accounts of these conflicts have tended to revolve around the term "service class". The movement of ex-urbanites into rural areas is nothing new. Indeed Pahl (1965) highlighted this phenomenon in the mid-1960s, but it has much deeper roots. By the late Victorian period, for example it was fashionable for London's stockbrokers to live in Surrey and for Manchester's businessmen to have villas in the Lake District. During the past 20 years, however, the process seems to have become ubiquitous, no longer confined to those areas around major cities, and to have become recognized as a widespread national and international trend (e.g. Champion 1989, Champion et al. 1989, Fuguitt 1985). Recent explanations for this population movement have tended to concentrate on the rôle of the so-called "service class", a group or "class" situated between capital and labour that broadly speaking consists of managers and professionals (Cloke & Thrift 1990). This group has expanded in numbers during the postwar period, and particularly rapidly since 1970, partly as a result of the shift in employment from manufacturing to services in the economy as a whole (Thrift 1987a). This class has a distinctive life style linked to certain consumption practices, such as interchangeable housing locations born out of a high geographical mobility (Savage et al. 1992), and it has exploited this mobility by seeking out housing in rural locations. According to Thrift (1987b: 78), the rôle of the service class is thus pivotal to rural change because it has "causal powers" that enable it to "take the lead" in creating new rural spaces. We do not address this analysis here (but see Murdoch & Marsden 1991) except to note that this account of the distribution of population and, to a lesser degree, of employment does bring production and consumption relations together in an especially forceful way.

We would, however, caution against the assumption that it is possible to read off the social consequences arising from the movement of ex-urban, middle-class residents into rural areas, partly because of differences in local historical experience and partly because the ways in which population change is regulated are quite variable. For example, new patterns of exclusivity arise not only because access to housing is mediated through the price mechanism in the private housing market and the availability of private transport, but also because the restrictiveness of the planning system in allocating additional land for housing has varied (Pahl 1965, Connell 1974, Hall et al. 1973, Newby et al. 1978). The earlier planning policies of key settlements and urban containment in the UK were premised on the need to minimize the loss of agricultural land and to consolidate new housing in order to keep public infrastructure costs low. However, continuing growth in real incomes, and home and car ownership, have intensified and spread the pressures on rural housing markets. These pressures have in turn undermined the settlement hierarchy on which rural planning policies were based. As a result, localities have increasingly had to compete with one another either to attract or ward off private and public investment. But what counts locally as wanted or unwanted development has also become a matter of growing contention. Thus the politics of place have become central to the pattern of private and public investment, within the broader constraints of macro-economic policy, the relocation of jobs, and developments in transport and communications (Johnston 1991).

While the economic fortunes of places are subject to these competitive pressures, their identity is also the outcome of processes of social differentiation, shaped in particular by residential mobility. One consequence is that prosperous and mobile social groups often come to share the same space – say, a village or an inner-city neighbourhood – with others who are tied there by occupation, residence, low income or kinship. This provides the social conditions for new representations of locality (Ch. 6) and new patterns of inequality arising from differential access to housing and labour markets, and to private and public services. Tensions may ensue from the different visions and expectations people have about the same place. Local government is frequently the focus of these tensions.

These changes are opening up new opportunities for capital in the exploitation and reproduction of markets and livelihoods in rural areas. Best known are the opportunities arising from the decline in state support for agriculture. On the one hand, the decline is strengthening the position of those who argue that there is a "surplus" of agricultural land and that farmland therefore generally needs less protection against developers. This is underpinned by the decreasing demands for agricultural land arising from the technologically induced growth in output. It is suggested that the exploitation of rural land for gravel extraction, golf courses and small workshops, for example, is now pivotal to providing the means by which accumulation can occur and (more liberalized) market demands can be satisfied (Ch. 5). On the other hand, in the more rural parts of continental Europe the primary concern is one of rapid rural depopulation, abandonment of farmland and the loss of a valued rural society. In these areas, the processes of commoditization, to which we now turn, are much less in evidence, reflecting the uneven form and pattern of post-agricultural development.

Commoditization processes: economic and political dimensions

The term commoditization describes the extension of markets to new spheres of activity or, more usually in advanced economies, the superimposition of new types of market relation (Long & van der Ploeg 1988). While this process may be ever more ubiquitous, it is also unstable. Capital seeks to transform new and existing use values into exchange values and simultaneously to develop new needs and markets. Successive bouts of commoditization – aided, for instance, by the privatization of services, utilities and public housing and efforts to liberalize rural planning – transform use values and outdated exchange values, and are thus central to conditioning the rate and direction of rural change. Land – in the ways it is owned, occupied, used and viewed for prospective development – provides a particularly effective illustration of the processes of commoditization (Ch. 4), revealing how different combinations of use and exchange value emerge in individual rural localities and how these change through time. The pressure to turn use values into exchange values has been especially intense under recent neoliberal policies. The nation state, through its twin programmes of deregulation and the progressive privatization of public services, has sought to construct new needs and markets, and even to make new commodities or recreate old ones in new forms. For example, many traditional agricultural buildings became redundant during the 1960s and 1970s as new machinery and equipment demanded larger, purpose-built accommodation. At that time, farmers came to see them as a financial liability. But by the 1980s, with increasing demand for distinctive rural housing and workshop premises, and state policy encouraging a revision of planning controls, such buildings began to assume new exchange values. These values were being socially and politically constructed at both the macro and local level, allowing considerable windfall gains to farmers as developers, even if these gains were actively opposed by other groups in the countryside.

The finance sector, particularly in its multidimensional provision of credit, is central to the processes of commoditization, not only as part of the general development of capitalism but specifically, today, as a function of the use of interest rates, by the state, as one of its key tools of macroeconomic management. The historically high real interest rate now being experienced in the UK has had major social consequences in a series of markets, as shown by the rapid rise in the number of repossessions in the housing market and bankruptcies among small firms. Banking capital is also adapting to, and moulding, the declining fortunes of a heavily indebted farming industry by extending and making more precise the conditions upon which loans are granted. The signs are that the rôle of the banks in agriculture will be increasingly focused upon short-term priorities and expedients designed to strengthen their control over creditors and to encourage the intensification of production. Moreover, in the UK, current policy moves towards the "diversification" and "extensification" of the farming economy, combined with a commitment in government to an "enterprise culture", will also increase the rôle of banks in directing rural change within and beyond agriculture. The rise of an indebted society represents, in itself, a measure of the growing extent to which households (as well as firms) are dependent upon commoditized exchange values. Enjoyment of much of the contemporary countryside is not free and the finance sector provides the wherewithal by which it can be selectively consumed.

In a somewhat different direction, developments in state policy designed to reduce agricultural production (e.g. set-aside policy, and milk and potato quotas) have also tended to develop new land-dependent commodities both within agriculture and beyond, for example, through the transferability of product quotas (Cox et al. 1988). Even policies designed to protect the environment, such as Environmentally Sensitive Areas (ESAs), succeed only by placing on pieces of land a politically constructed, commoditized value designed to compensate farmers for the loss of exchange value in agricultural commodity markets arising from the designation, a value that in turn derives from political negotiations in the context of the Common Agricultural Policy (CAP) of the EC (Ch. 4). Government has seen this compensatory device as a means of placating farmers in the face of their falling incomes from food production. But as the fall in incomes arises largely from a reduction in state-supported commodity prices, the scope to compensate farmers is also being diminished. In consequence, government has encouraged moves to commoditize an ever-widening range of land-based activities and to orient these towards non-agricultural markets (Countryside Review Panel 1987, MAFF 1987, Major 1992, NFU 1992). In this way, the diversification of the farming economy has been oriented towards the re-use of surplus property rather than surplus labour, and to encourage the holders of property rights to enter a wide range of more unpredictable markets in which they have, for the most part, less expertise. The need for farmers in particular to acquire marketing and property development skills, in addition to their husbandry and entrepreneurial talents, has never been more apparent.

The attempt to exploit rural space by opening up new markets is far from being a smooth or even process. It leads to acute conflicts between, for instance, the protection of collective consumption-oriented use values (e.g. public recreational access to meadows, woods, viewpoints, etc.) and the attempted imposition of private, production-oriented exchange values (mineral extraction, house-building) because it adjusts the social basis of entry (access) from ones of customary rights (both public and private, legal and informal) towards ones based upon economic power. Clearly, the rôle adopted by the state at all levels is crucial in these struggles, but a consistent approach cannot be assumed. For example, there are considerable differences in outlook between rural authorities in northern England, seeking to attract new employment, and those in the south where a stronger emphasis is placed on rural conservation and environmental protection. As will be evident from this discussion, it is not simply the replacing of exchange values for use values that is significant, but also the social processes underlying these changes in the definition and trading of goods and services, and their realization through political action at the local, national and international scales.

Representation

In seeking to incorporate the political basis of action into our analysis we need to focus upon the rôle of actors in these commoditization processes, and this is best achieved by focusing on the representation of interests. As we have already suggested, representational activity provides an important mechanism for reproducing the rural domain as both a physical and an ideological entity. For instance, recreating scarcity value and the authenticity of rural living depends upon the social maintenance of exclusivity itself, which in turn depends upon social representations in local and national networks of power. Localities themselves, as we will discuss in Chapter 6, are not only spatial entities; they also have trajectories through time. Their dynamism springs from the continuous reconstruction of the local social system. A shift towards understanding the locality as a social construct means that we must examine how such constructions or representations come into being.

If representations are to be regularly utilized as a means of achieving or resisting change, they must exhibit some level of stability and enable people to make sense of their worlds. Representations are thus utilized in specific situations, such as the land development process (Ch. 7), but they also allow local action to be linked to wider economic and political processes that may lie beyond the locality but be of considerable influence upon it. In this sense, commoditization may in part be a representational process. For instance, the development of industrial units in rural areas, or the use of agricultural land for afforestation, is actively pursued by certain development-oriented interests. Arguments are marshalled concerning the need for diversified economic activity in the countryside, and then promoted by, *inter alia*, linking them to aspects of central government policy or to the activities of public agencies. The countryside is represented as a fresh economic space, an opportunity for capital investment that may, in the process, create new exchange values. Because the values these interests represent may be challenged or even reversed, the concept of representation provides a means of examining the dynamics of change. All representations are subject to reconstruction and ultimate rejection, however dominant particular ideologies appear in the short term. Like commoditization, the processes of representation are subject to continual struggle and re-negotiation.

Of particular concern to us is how actors construct their interests, how they seek to represent them and how effective they are in achieving their proclaimed ends (Ch. 6). Frequently, for example, representors claim the right to act and speak for the represented, but a representation cannot capture all there is to be represented. There is always more that could be said. As Knorr-Cetina (1988: 43) argues:

> The representing and the represented are best considered to pertain to different, co-existing realities regardless of the claim of one to speak and stand for the other. The case can be made that claims to represent are at the same time political strategies, political topics and resources in the power struggles of everyday life.

Such "power struggles of everyday life" are fought out in particular local and national contexts. Those doing the representing have varying economic, social and knowledge resources at their disposal in their attempts to ensure that others share their representation. Those who share the same representations seek to reinforce them, while those who do not may contest them (Long & van der Ploeg 1989). Linking to the previous section, the development and reproduction of representations are central to the pace and direction of commoditization and thus to the macroeconomic restructuring of production and consumption; and in empirical terms it is important to develop means of following how representations flow along networks (for instance those associated with institutional, economic or political agencies) and how such networks are able to effect change (Ch. 6).

Mormont (1987, 1990) has been one of the few researchers to adopt this approach to the study of rural change. He analyzes the struggles over representations of nature and rurality, noting how the representations of competing groups "shape" social space. In discussing rurality he argues (1990: 32):

> we should start from the hypothesis that the rural/urban opposition is socially constructed and that the rural exists primarily as a representation serving to analyze both the social and space – or rather to analyze the social while defining space. The fact that it is a constructed representation and not an ascertained reality does not deplete a sociology of the rural of the subject. Its subject may be

31

defined as the set of processes through which agents construct a vision of the rural suited to their circumstances, define themselves in relation to prevailing social cleavages, and thereby find identity, and through identity, make common sense.

This perspective allows a local dynamic to interact continuously with the more global processes that surround the local. What Mormont fails fully to appreciate, however, is the importance of the ways in which the social construction of the rural is constituted out of *competing* representations. These competing representations are not free social relations but negotiated by networks of actors, linked through relations of power, and able to utilize differing sets of resources – material, cultural and symbolic. In the rural arena, such resources are associated, to varying degrees, with the ownership and occupancy of land (Ch. 4), and thus the relative power of representations will partly be conditioned by the representors' control of land based resources. Their legitimacy will depend upon their ability to construct compelling arguments based on the assertion of rights in real property. This raises the related questions of how we conceptualize property rights and interests in land, and how any particular society assesses the significance and distribution of property rights in its allocative decisions.

Reintegrating property relations

Everyone has interests in land, at least as consumers if not as producers of its goods and services. It is not necessary to hold private (legal) rights to have an interest; social interests are not reducible to landed interests, even though the former are mediated through the latter. More properly, the concept of interest in land refers to the relationship – material, ideological, symbolic – between an actor and landed property. The interest arises from the objectives of the actor concerned (income generation, continuity of occupance, protection of the environment, etc.) and is reflected to varying degrees in the rights being asserted. Rights are thus those interest claims that enjoy some legal, moral or social sanction. In these terms, the notion of property as a bundle of rights is a powerful one, provided the rights are understood as embodying social and economic relationships, such as those of power, custom or kinship. Their existence is dependent upon the changing character of the sociopolitical system that regulates them, and are of no purpose unless their authority can be sustained. As Harrison (1987: 37–8) says:

> depending upon the shifting sands of recognition, property "rights" would come and go. Of course this is what happens in practice, emergent claims being placed on the agenda of current politics. At any time, however, there are degrees of recognition of rights claims. Some are backed fully by law, others by administrative custom, and others only by assertions about morality. Even the most formally recognized rights may clash, for legal frameworks are rarely fixed or absolutely clear. Furthermore, some claims are not often expressed formally, yet seem implicit in widespread material conflicts of interest, as "submerged" rights claims to which parts of the political system may be under pressure to respond.

For rights to be enjoyed, holders need to be able to capture the benefits. In the absence of such exclusivity, the means of accumulation can dissipate rapidly unless an acceptable form of compensation can be negotiated for loss of right. This issue is central to the conflict between, say, the private rights held by farmers and the collective public benefits from the countryside sought by environmentalists. As Bromley (1991: 3) emphasizes: "environmental policy is nothing if not a dispute over the putative rights structure that gives protection to mutually exclusive uses of certain environmental resources".

The common law in the UK strongly supports exclusivity. As Newby et al. (1978: 337) argue:

> legally constituted property rights are concerned with specifying the boundaries on access or exclusivity, they are not concerned with specifying when or how the various control, benefit and alienation functions are achieved, except within very broad regulatory limits. *Access* is stipulated by law, while *use* or *function* is left virtually unfettered. The law is thus prescriptive with regard to exclusivity rights but permissive with regard to property functions.

This statement may be regarded as too sweeping as the exercise of property functions is subject to widespread modification by statute, including the operation of the planning system. Nonetheless, it represents the starting point from which landed power is drawn. In broad terms there are three key areas of rights that are formally recognized in law: the right to transfer (*owner rights*), the right to use (*user rights*,including existing, new and antici-pated uses), and the right to exclude others from the property owned (*occupier rights*). Where all the rights are held by one individual or legal entity, the land may be said to be in freehold ownership.

Freehold ownership incorporates a wide range of rights and a high level of divisibility. We will suggest that the latter has not arisen by accident, but is a response to the requirements of the different phases of capitalism. Divisibility facilitates the commodi-tized exploitation of property in general and the penetration of traditional rural landholding arrangements by industrial and finance capital in particular. The ability to subdivide encourages the multiple use of property as a means of maximizing its value.

Similarly, the use and development of land regularly incorpor-ates the short-term ownership of at least some of its rights. This is well illustrated by developers who may not, initially, purchase the freehold, but negotiate an option or conditional contract with landowners. Part of the land may then be sold to another builder, with the freehold being divided up and sold to house-buyers. Thus, a given distribution of rights will evoke quite differing sets of expectations in exploitable interest between actors. In a commoditized market, such expectations will be reflected in market valuations and will form the basis of fictitious capital against which owners may borrow (Harvey 1982). At the same time, flexibility in the ownership of rights facilitates the switching of capital, the reduction of entrepreneurial risk and the striking of compromises between oppositional interests, all designed to aid the process of capital accumulation (Whatmore et al. 1990). Two methodological issues arise. How are interests in land identified, including the relationship between interests and the behaviour of actors? And how are interests in land promoted and represented?

In their discussion of the first of these issues, Healey et al. (1988, Ch. 7) identify two opposed schools of thought. One, which they term positivist, argues that action can be taken as an expression or approximation of interest. Interest can thus be established empirically through the observation of behaviour. This position has been challenged on several grounds: most tellingly, for assuming that inaction is tacit acceptance of the status quo, thus ignoring the possibility that it is the conscious outcome of contradictory interests; and that actors, as if indepen-dent of their social context, are able to determine their interests subjectively and freely. The other school, which they characterize as Marxist, starts from the reverse position in which individual interests are reduced to class interests objectively determined by position within the class structure. But this school too has been roundly criticized on the grounds that other cleavages in society (race, gender, religion) may be of comparable significance to class in influencing both interest and action (Lash & Urry 1987). Similarly, reducing the interests of actors and institutions in land to material ones denies experience regularly reported in which, *inter alia*, commitments to stewardship, environmental protection or family continuity cut across financial considerations (e.g. Salamon 1980, Hamnett 1987). While material interests have undoubtedly been promoted by the commoditization of the countryside, the range of interests pursued by both producers and consumers varies from place to place, and it is essential to incorporate notions of local resistance and representation into our analyses of action. In sum, there is a growing range of social forces generating more complex sets of material and non-material interests in land that extend from, and indeed go beyond, conventional class categories.

The second issue concerns the manner in which representative bodies associated with land promote the interests of those for whom they claim to speak. There are many such bodies (e.g. in Britain, National Farmers' Union (NFU), Country Landowners' Association (CLA), Council for the Protection of Rural England (CPRE), Timber Growers' Association (TGA), Housebuilders' Federation (HBF), etc.). For them, a conflict arises between the wish to encompass as large a constituency as possible and the maintenance of a coherent position. Membership size is a significant issue for any group claiming representative status. The more inclusive a group is of a particular population, the greater the authority it has in defining and promoting the interests of its constituency, encouraging other authoritative organizations to recognize its claims to representative status and discouraging the formation of rival groups. At the same time, government increasingly looks to those groups accorded consultative status not only to represent but also to aggregate interests. Indeed, a feature of the more longstanding and successful bodies is their ability to manage and contain differences among their members – for example, within the NFU, between corn and horn, and between tenant and owner-occupied farmers. In practice, the definition of a group's interest (whether food production, nature conservation or whatever) and how it is to be represented is often determined by its ruling elites, whose continued ascendancy usually depends on their ability to maintain unity and contain any internal dissent.

In the UK, many representative groups or organizations assert interests in land, but only a few have either a direct interest as owners of private property rights (e.g. National Trust, Woodland Trust, Royal Society for the Protection of Birds), or make property rights issues their *raison d'être* (e.g. CLA). Rather more promote their interests indirectly through the support of particular forms of production or consumption, but do so on behalf of a membership with private rights to rural property, such as the NFU, CPRE or HBF. Some, such as the NFU, have professional departments devoted to defending members' property rights through, for example, participation in planning inquiries. Others are concerned to establish new collective rights, and to expand existing ones. The Ramblers' Association endeavours to extend public access to the countryside, the benefits of which are, by definition, not exclusive to its membership. Finally, as an illustration of how things change, the CLA and NFU were established at the beginning of this century as oppositional property interests, i.e. landed versus productive (tenant) capital. During the middle decades of the century, the significance of this opposition became increasingly secondary to their common interest in the promotion of state support for expanded agricultural production. It was also undermined by the rapid growth of owner-occupation, leading to an increasing cross-membership between the two bodies (Ch. 4). Significantly, the crisis in productivist policies in recent years has seen the re-emergence of certain tensions, with arguments over which fraction of capital should benefit from non-production-oriented state payments to agriculture (for example, for conservation, access and the setting aside of land) and over whether the reform of agricultural policy should emphasize farm income support or the freeing-up of landed assets.

Conclusion: restructuring, regulation and rurality

We have attempted in this chapter to outline some key areas for conceptual development in the study of rural change, during a period when capitalist economies are evolving revised forms of accumulation, consumption and regulation. These arguments have arisen out of a dissatisfaction with the current restructuring literatures whereby a heavy emphasis is placed on economic and productionist logics of development that are assumed to be applied "top down" on local social relations. Nonetheless, we accept that current restructuring processes represents a significant break with the past (Martin 1989) even though a real or conceptual coherence about how this should be expressed has not yet emerged. Whereas Fordism could plausibly be interpreted in terms of the diffusion of the American model of mass production and mass consumption to other national economies, there is no revised growth model of corresponding status. A focus upon the significance of international finance and industrial capital, or more specific notions concerning flexible specialization and product differentiation, are, of course, key features; but they do not as yet fit together to provide a clear picture of the links between local, national and international forms of production, consumption and regulation. Neither, as Hirst & Zeitlin (1991) point out, do they sufficiently incorporate the variability and salience of the processes of social organization involved in these spheres.

Despite the appeal of the integrative nature of the regulationist approach to restructuring, it is not exempt from critical assessment. We do not wish to undermine its general relevance, but it is necessary to build and apply a revised set of concepts capable of allowing greater insights into the changing position of rural areas.

First, as our discussion demonstrates, debate has been pitched at a highly abstract level. This has led to many fewer attempts to develop analytical concepts directed towards relating specific events to theory. A major problem has arisen concerning the best ways empirically to examine restructuring processes. Second, and related to this point, has been the reluctance to examine diverse regional and local power structures and to assess how these are embedded in national and international economic and political frameworks. Although much of the British restructuring literature has focused on relating global processes to local "responses", this has been undertaken in a largely economistic and somewhat mechanical way. As our discussion of commoditization, representation and property rights suggests, local systems are capable of shaping development in unique ways. They can influence the overall logic of capital penetration and state intervention. Moreover, although the regulationist approach does identify the significance of the state in moulding social and economic relationships, the understanding of these relationships, particularly their dynamic character, is seldom pursued through concrete analysis. As Jessop (1990: 24) argues:

unless one examines the mediation of regulation in and through *specific social practices and forces*, regulation will either go unexplained or will be explained in terms of "speculative" structuralist categories. Yet the regulation approach in all its guises was developed precisely in order to overcome structuralism as well as mechanized theories of general economic equilibrium.

As we shall outline in the following chapters, with reference to the UK, the centrality of regulation needs to be viewed as both a cause and a product of contested social space, and of class and property relations. This involves the study of particular *contextualized* power networks whereby the political and the economically exploitative come together and are reproduced. Only by focusing on specific national and local case studies, and by drawing on the type of middle-level concepts outlined here, can theoretical and methodological progress be readily made. Similarly, analysis must take into account the globalization of industrial and finance capital, and, possibly, greater evidence of transnational forms of governance (McMichael & Myhre 1991). Whether such tendencies will increasingly condition the exploitative potentialities of national and local economies remains to be seen. Nonetheless, whatever their significance, their consequences can only be understood by reference to grounded social contexts, revealing the capacities of national and local governments and quasi-governmental agencies and institutions to direct, resist and legitimate the actions of strategic capital.

We have argued here for a set of middle range concepts that we believe move us beyond some of the broad theoretical arguments dominating much of the restructuring literature. These concepts derive from a broad review of this literature and, as we see it, their value when applied to the reorganization of rural space. In particular, the debates surrounding such terms as Fordism, Post-Fordism, flexible specialization, and the service class alert us to the importance of the changing relationships between production and consumption. They force us to confront the ways in which rural space is being commoditized as a result of the activities of new forms of capital, the redirection of state agencies and the changing interests of consumer groups. It will also be evident that the commoditization process does not refer to the simple (almost "magical") unfolding of market forces; it entails struggle, as some actors impose upon others new representations of rural space.

How or why certain actors are able to dominate is linked to

their access to and use of resources, and we have identified property rights as a particularly important source of power in the rural arena. Even so, how property rights are utilized in the processes of commoditization and representation is by no means fixed and unambiguous; such rights have to be continually upheld in a multiplicity of situations where their meaning and scope may be under constant challenge. Commoditization presumes the assertion of particular property rights regarding transfer, use and exclusion. The definition and enforcement of such rights are indeed essential to the proper functioning of any market. Here the state plays a crucial rôle upholding private property rights but also periodically modifies them to accommodate forces of restructuring. Property rights and the way they are maintained and modified by the state are thus instruments of regulation.

The value of these concepts can only be reaffirmed through the analysis of real instances, and to sustain our position we propose, at this point in the book, to examine a case that is well known to us. Only later (in Chs 6 and 7) will we return to the methodological implications of applying our conceptual approach to the study of specific localities. The case study should be viewed as an exemplar of our ideas rather than an exhaustive analysis of a particular set of circumstances.

The following three chapters thus contain an assessment of the nature of rural change in the UK over the past century. Each chapter acts as a different prism through which change may be examined. In Chapters 3 and 4 we adopt an historical perspective, reflecting the importance we attribute to inertia and continuity to any understanding of contemporary rural social relations. In contrast, Chapter 5 is based quite deliberately on the unfolding events of the 1980s, the decade during which the tendencies towards deregulation and post-productivism that we identified earlier could be expected to have their greatest impact on rural space. It would, however, lose much of its insight without the historical analysis of the preceding two chapters.

In Chapter 3 we focus upon the changing nature of national regulation since the mid-19th century and its implications for production and consumption in rural areas. The analysis is primarily directed towards the UK's changing position in the global political economy, what this has meant for the farming industry and how the interests that support agriculture have at different times sought to represent their position. Chapter 4 parallels this account by tracing changes over a similar period in the pattern of rural property rights as key instruments of regulation. These changes provide sharp insights into rural power structures and revised representations of the rural. Processes of commoditization are revealed by the decline in the ideology of stewardship and consumption that once dominated landowner-ship, brought about by the state's promotion of a productivist agriculture from the mid-1930s onwards. The growing commoditization of consumption interests in land is also partly the result of state policy towards the support of agriculture and the containment of urban growth, and these tendencies form the context for Chapter 5. In this chapter we examine new forms of representation and how they have influenced the functioning of the local planning system during the 1980s. The planning system is seen to act as a key arena of struggle between competing interests, and its significance to the uneven emergence of a post-productivist countryside has been emphasized by indecision in central government over how far deregulation, as an ideological goal, should be pursued within the rural domain.

CHAPTER 3

Agricultural regulation and the development of rural Britain

Introduction

The full significance of the debates surrounding the nature of rural restructuring outlined in Chapter 2 only becomes apparent when grounded within particular contexts. The context of concern here is the postwar development of the rural UK and the next three chapters are devoted to our assessment of it.

In spite of the increasing internationalization of economic forces, political institutions and technological trends, historically specific local and national conditions and institutions remain potent influences on the course of events. They affect the range and priority of public concerns, the nature of policy responses, and the means of their implementation. The distinctiveness of the UK's circumstances, especially its institutions, cultural perspectives on the countryside and dominant economic interests, needs to be set within an appropriate historical frame. The temptation is to restrict attention to the past half century, the period during which the state has exerted a profound effect on rural development. But that would underestimate key aspects of continuity that have moulded postwar rural policy. At the risk of overgeneralizing, some distinctive features can be identified that are of much longer standing. They have been crucial to the form of state action adopted since the 1940s, and its consequences.

First, in spite of the "backwardness" of some peripheral regions (Carter 1979), capitalist relations of production were widely established in agriculture at a very early stage, and before the flowering of industrial capitalism. By the late 18th century, many parts of the UK already possessed a fully commercialized agriculture. It was self-consciously innovative and oriented towards the expanding market in food commodities presented by the burgeoning urban population. A capitalist labour process also rendered labour exceptionally mobile. Combined with a laissez-faire policy towards domestic agriculture from the 19th century onwards, these conditions led to a technically efficient farming industry characterized by large farms employing an ever smaller proportion of the nation's workforce.

Secondly, the dominant factory system in urban areas eliminated long-established cottage industries in the Victorian countryside, leaving a rural population singularly dependent upon agriculture (Mingay 1981). On its own, the countryside could not provide the range of employment opportunities necessary to sustain a dynamic, confident and broadly based rural society. At the beginning of the 19th century, one-fifth of the population lived in towns; by mid-century, the urban population had overtaken the rural; and by the end of the century, four-fifths of the population was urban-based. Urbanization and industrialization thus produced a spatially polarized as well as a nationally integrated geography. The bulk of the population and commercial and industrial activities were concentrated in the cities, while rural areas became dominated by a technically progressive, market-oriented agriculture.

Thirdly, unlike much of the remainder of Europe, there has been no peasantry, or large class of smallholders with extended kinship and community links, to be weaned off the land during the 20th century; for this reason, political resistance to the fluctuating economic fortunes of the countryside has been relatively muted. On the contrary, the absence of a substantial peasantry and the political clout it could mobilize has contributed to a national perspective on the countryside that is strongly urban or consumer-led. The influence of this perspective on public policy has been profound, and not without its contradictions. In the area of agricultural production, a broadly economistic view has prevailed. The farming community has not been valued so much for itself as for the food it could produce, and then at the lowest reasonable cost. As we will argue, rural policy has been a minor adjunct of agricultural policy, and this in turn has been subordinate to other concerns: food supply, national competitiveness and international trade.

The 1880s and 1890s saw a major agricultural slump, particularly in the arable counties, as cheap food flooded into the UK from overseas. The prime cause was the agricultural development of the interior of North America, Australia and temperate South America, which was linked to the decline in oceanic and railway freight rates, and the subsequent development of refrigerated storage and transport. The exploitation of vast, virgin land resources enabled extremely cheap production, made available for export through the extension of railways and inland waterway systems. Whereas other European countries moved to protect agriculture by introducing tariffs, the UK remained doggedly committed to free trade. It was believed that in order to maintain the UK's manufacturing competitiveness, food had to be obtained from the cheapest sources, thus keeping down the level of industrial wages (Tracy 1982). To the UK's urban majority, free trade meant cheap food and successive proposals for tariff reform consistently failed to win electoral support. Command of the seas, and the dependence of many exporters of primary commodities on the British market, was sufficient to ensure food security and to allow a fall in the level of domestic self-sufficiency. In a sense, agricultural prosperity and the rural population became victims of the increasingly fierce trade competition between the UK and its industrial rivals. What had been a vibrant sector for investment and profit before the 1870s became an economic backwater. Land went out of arable cultivation, the rural population declined sharply and many village communities became depressed and stagnant.

It was not until the late 1930s that the beginnings of a sustained recovery emerged. This laid the basis for an unprecedented period of prosperity for British agriculture during and after the Second World War. In the 1980s, depressed conditions returned. But the tremendous increase in farm labour productivity in the intervening period, and the growth of the service sector and some manufacturing in rural regions, had greatly reduced the countryside's economic dependency on farming. Despite this decoupling, indeed partly because of it, the impact of the current agricultural depression has been profound in catalyzing a broader redefinition of the social functions of rural space, away from agricultural production towards a greater emphasis on its consumption rôle. This is part of a broader transformation from a highly specialized space devoted almost exclusively to primary production to a socially and economically much more diverse "post-industrial" countryside.

In Chapter 2, the notion of regulation was introduced to signify the negotiated order within a society between production and consumption. Broad shifts in the function of rural space follow as a consequence of changes to that overall order. Specifically, we would argue that state management of the countryside has not been directed specifically towards the needs of rural inhabitants, but has traditionally been a function of the regulation of the *agricultural* economy and this in turn has been a function of the UK's food policy and its international trade and economic relations. The latter have undergone a profound transformation during the past 100 years or so, and it is against the backcloth of the UK's changing position in the global political economy that the state has interpreted the nation's food security.

This chapter identifies two broad phases in the contemporary regulation of these relations; an *Imperial food order*, dominated by the UK and lasting from the 1860s to the 1930s; and the postwar *Atlanticist food order*, dominated by the USA. In the first phase, the rôle of agriculture was as a source of cheap "wage goods" during a period in which the real price of food remained a primary determinant of the cost of the reproduction of labour. Extensive development of agriculture was encouraged largely through the vast extension of the frontiers of agricultural production and trade through the agency of colonialism and the imposition of free trade. In this way, international commodity markets were established and rural areas, at home and overseas, became tied into them. In the second phase, a more intensive development of agriculture was encouraged as part of a shift towards a mass consumption economy and greater dependency on domestic food supply, for strategic reasons linked to the UK's declining military and economic strength. With wages linked to productivity gains via collective bargaining, the industrial labour force was no longer regarded just as a cost to capital but also as an expanding market for manufactured consumer goods and food products. In this context active government support for domestic agriculture was oriented towards ensuring relatively cheap, abundant and secure food supplies. That support facilitated the expansion of domestic agriculture and its industrialization became a focus of accumulation.

This chapter briefly discusses the rôle and fate of British agriculture and rural regions in the Imperial food order before examining in greater detail how, in the postwar UK as part of the Atlanticist food order, a productivist regime of agro-industrial accumulation was established that not only regulated the farming industry but was also the dominant factor in the socioeconomic management of the countryside. The chapter will analyze the UK's distinctive rôle in international food trade and the key elements in its productivist regime, including the special significance of land regulation. The general breakdown of the productivist international food order in the 1970s led to the international farm crisis of the 1980s. The chapter concludes, therefore, with an analysis of the specific British response to the crisis, focusing upon the alternative land-use debate.

The Imperial food order, 1860s–1930s

By the beginning of the 20th century, the UK was heavily dependent on imported food. Not only was a great variety of tropical products imported, but also temperate products, including three-quarters of the country's consumption of wheat and cheese, and half its meat. By developing manufacturing industry and allowing agriculture to shrink through a policy of free trade, the UK had become by far the world's largest importer (and processor and re-exporter) of food and raw materials. For many countries, the UK was the prime market for key exports: for example, for Russian and American wheat, Argentine meat, and Danish butter and cheese. The British market was vital for many Empire products: for example, Australian butter, wool, wine, beef, dried and canned fruits, sugar and lamb; New Zealand butter, cheese, lamb and mutton; Rhodesian tobacco; West Indian and Mauritian sugar; Ghanaian cocoa; and Malayan rubber. These industries had been established with the British market in mind, leading in some cases to a marked dependence. For example, by the 1930s, 15% of total Australian exports and no less than 70% of those from New Zealand consisted of meat, butter and cheese bound for the UK (Drummond 1972: 221).

These developments came about as a consequence of the UK's approach to international trade and the development of its Empire, and rested on its leadership in industry and commerce. That leadership had initially been built upon traditional colonial relations that had provided protected markets for trade in luxury goods and cheap manufactures. But it was transformed into a position of dominance through the expansion of this trade, and the opening up of similar relations with other countries, in association with rapid British industrialization. The consequence was a marked shift in the UK's international trade towards the import of primary goods and the export of infrastructural and capital goods. The British Empire was thus organized around the geographical "separation of agricultural and manufacturing sectors as poles of imperial exchange" (Friedmann & McMichael 1989: 96).

The UK's economic ascendancy made its manufacturers, traders and politicians particularly receptive to the free trade arguments of Adam Smith and his disciples. Through a series of tariff reductions and removals, beginning in 1822 and reaching its most controversial point in the repeal of the Corn Laws in 1846, the UK became a free-trade nation. Formerly protected colonial markets were also opened up to world trade. A free-trade regime was imposed through a combination of naval power and diplomacy, industrial and commercial might, and an efficient financial infrastructure based in a reorganized London discount market with sterling as the international currency (McMichael 1985). The result was the establishment for the first time of a unified, price-regulated world market.

The UK's global hegemony induced a response on the part of both its continental rivals and certain settler societies, including the USA, that eventually yielded alternative models of economic and political development. In striving to construct their own national economies, these states had to work within an international framework in which they were obliged to compete on the UK's terms, while simultaneously seeking to regulate the effects of closer integration into the world market and to promote their own economic development. The result was the extensive progress of twin movements: for political and commercial liberalism on the one hand and economic nationalism on the other.

The interaction of these twin movements, in a context of fluctuations in world trade, produced contradictory developments in the regulation of national markets. Thus the Anglo-French Treaty of Commerce of 1860 was a major step in bringing down tariffs across Europe. The duty reductions and abolitions promised to France were extended by the UK to all other countries, and for its part, the French government used the Treaty as the cornerstone of a series of subsequent agreements with most other European countries. But this period of European trade liberalization proved shortlived, and from the late 1870s onwards most countries began to return to protectionism (Tracy 1982). The causes included the growth of nationalism, particularly in the wake of the Franco-German war of 1870 and a long-run trade depression. The depression was caused by overproduction both of Continental industrial goods and of cheaper grains from settler regions developed with British capital. Together they induced a downward trend in prices between 1873 and 1896 and a consequent squeeze on profits and wages (Hoffman 1932, Hobsbawm 1975). Agricultural opposition to the rising tide of cheap food imports thus coincided with growing demands from manufacturers to protect both traditional and infant industries from foreign competition. The result was a wave of protectionism as national governments responded to pressures from agricultural and industrial interests, with the actions of one country provoking countermeasures in others. In most cases, measures to protect agriculture were introduced concurrently with others to shelter growing industries behind tariff walls. The UK, though, held firmly to free trade, as did some smaller countries, notably Denmark and the Netherlands, that were heavily dependent on trade and access to the British market. But these other countries sought to restructure their agricultural sectors through expanding and improving the competitiveness of livestock production. The approach was based partly on the growing availability of cheap imported grains, but also on active support for farmers through technical education, provision of credit and encouragement to form cooperatives. The UK, by contrast, adhered to a laissez-faire approach, and adjustments were forced on its agriculture in the form of contraction and decline in the arable sector.

The peace that prevailed in North America and Europe for 40 years after the Franco-German war was conducive to a great expansion of trade. A series of exceptional harvests in North America in the late 1870s and early 1880s coincided with poor years in western Europe. Previously, higher prices had compensated European farmers for reduced harvests, but now imports rose sharply, prices fell and farmers suffered severe losses. From then on grain prices fell steadily. By the early 1890s, Canada was becoming a significant exporter, joining the USA and Russia, and the Suez Canal was facilitating imports from India and Australasia.

In its early stages, the agricultural depression affected mainly the grain market. A decline in the prices of livestock products did occur, but somewhat later as techniques of refrigeration began to be applied on a commercial scale. Shipments of frozen meat from America began in 1875 and from Australia in 1877. The drop in price, however, was not as sharp as for grains. This was because production was rising less rapidly while demand for livestock products was expanding sharply with improvements in living standards. The most important beneficiary of rising incomes was the liquid milk sector, with transport improvements allowing easier collection and distribution locally while its bulky and perishable nature protected producers from imports. Livestock producers, moreover, were cushioned from the fall in their prices by the much greater fall in the costs of feed grains. The inevitable consequence was a shift from crop production to livestock in European countries, particularly those that had not succumbed to protectionism. British agriculture was deeply affected by the depression: between 1871 and 1901 the agricultural population fell by over a fifth, or from 15 per cent to 8 per cent of an otherwise expanding national workforce.

Since the Second World War and the shift to an agricultural policy that has emphasized the protection of domestic production and the quest for self-sufficiency, the free trade and laissez-faire approach that prevailed previously has often been portrayed as reckless and misconceived. Not only has it been condemned as the cause of extensive rural decline and impoverishment, but also for undermining the country's food security and thereby rendering the UK dangerously exposed to a food blockade. For example, it has been estimated by Grigg (1989) that by 1914 about half the food supply was imported and on the eve of the Second World War this had grown to about two-thirds. Grigg's comment (p. 8) that "this system of provisioning the country, whereby cheap food was paid for by the export of manufactured goods, received a rude shock in the U-boat campaign" is not untypical.

It is important to realize, however, that British politicians and policy-makers had not been unaware of nor necessarily unreceptive to arguments concerning food security, but that such arguments were cast in a different strategic context in the 19th and early 20th centuries. For some of the Continental powers, the maintenance of a large rural population reflected not only the entrenched position of conservative social forces based in rural areas but also national concerns about food supplies and the need to maintain military reserves based on the peasant stock. For the UK such strategic considerations were framed against the backdrop of Empire. The economic development of the Empire had proceeded on the assumption that the UK would supply manufactures, capital and migrants, and the rest of the Empire would supply food and other primary products (McMichael 1985). The UK also looked to colonial and dominion governments to support the collective defence of the Empire. Indeed, the peopling of the Empire through white settlement was seen to be a means of ensuring the UK's own long-term military security as well as securing Imperial defence.

With such an extensive food-producing Empire, the UK's food security depended critically on its continued undisputed mastery of the seas. This in turn depended on such specific factors as the size of its shipping fleet and its shipbuilding capacity, as well as on general factors such as the nation's overall financial situation and its manufacturing strength, all of which could arguably be harmed by a policy of domestic agricultural self-sufficiency in peacetime. Furthermore, neglect of the UK's domestic agriculture did not imply any lack of interest in rural development, at least overseas. On the contrary, it was a central feature of Imperial policy to facilitate the exploitation of the primary resources of the Empire through the encouragement of emigration from the UK and the supply of capital on favourable terms. In return, Empire governments looked to secure preferential access to an expanding British market to absorb their growing primary production. The Finance Act of 1919 introduced the principle of Imperial Preferences, through which imported goods from the Empire received concessions on some customs duties. This further encouraged the expansion of primary production in the Empire for the British market. For example, in Australia, whole tracts of land were newly settled for fruit farming and grape growing.

By the interwar years, the question of the UK's dependence on imported food and raw materials and the reciprocal dependence of much of the Empire's primary exports on the British market had become central to the debate between free trade and tariff reform. This was a long-standing controversy in British politics which had until then been marked by the strong ascendancy of the free traders. But after 1919 the discussion took place in a changed international environment. The UK had emerged economically weakened from the First World War, and unemployment was a chronic problem. Important politicians in all three parties were still devoted to free trade, but the Conservative and Labour parties both included devotees of a protective and preferential tariff system reform, usually cast within an Imperial context. The lack of decisive popular support for tariff reform, particularly because of its likely adverse impact on food prices, was registered in the 1923 general election which Stanley Baldwin fought (and lost) for the incumbent Conservatives on a platform of economic protection linked to development of the Empire.

The 1929–33 slump changed attitudes. Even during the 1920s, it had become evident that some protectionist countries were booming, while the free-trade UK was recovering slowly and with difficulty. Throughout the interwar period, trade in manufacturing grew more slowly than output. As the slump deepened after 1929, more people came to support protective tariffs because it appeared ever more obvious that such tariffs would benefit them. The Conservative leaders, putting the 1923 experience to one side, were ready to articulate this change in public attitudes.

Many of them had been protectionist in outlook for decades, as were most of the landowners and industrialists who supported the party, and Labour leaders also increasingly suspected that tariffs would mean jobs, and that dearer food for the employed might be better than no food for the unemployed.

Between the 1930s and the years of postwar reconstruction, most Western industrialized nations introduced measures to counteract instability in domestic agricultural markets and to lift their rural areas out of the impoverished conditions of the interwar years. This was part of a more general change in the management of national economies in response to the Great Depression and its undermining of the classical liberal ortho-doxies of public finance. The mass unemployment and economic hardship of that period led to demands for extraordinary government actions on behalf of industrial workers, farmers, and other distressed groups. With the emergence of Keynesian eco-nomics, and shifts in political power that strengthened organized labour, states expanded their functions as the active agents of societal welfare through a synthesis of social spending and macroeconomic management (Peterson 1990). Although the UK was the home of free trade, it did not stand apart from this movement. The deep depression in both industry and agriculture in the 1920s and early 1930s made some protection for farmers politically possible at last.

In 1924, farmers were exempted from local rates (property taxes) and the following year a subsidy on home-grown sugar was introduced. Legislation followed to reorganize the marketing of milk, potatoes and hops. In 1932, more overt protection was offered to arable farmers with the introduction of a deficiency payment for wheat. A subsidy on fat cattle was introduced in 1934, and in 1937 the Ministry of Agriculture began to subsidize the use of lime. Despite this growing battery of supports, the scale of intervention remained modest costing the British govern-ment no more than 5 per cent of the value of gross agricultural output in 1937–8 (Whetham 1978). In part, this reflected the continued strength of free trade sentiments. The Labour Party certainly questioned the need to produce more wheat domesti-cally when it could be imported at a lower price (Williams 1935). In an influential expression of opinion, Astor & Rowntree (1938) likewise argued that to expand home production would be very expensive for government and the consumer, and the resulting reduction in imports would harm foreign relations.

In introducing measures to support farmers, the government

was indeed anxious not to provoke countermeasures that might jeopardise industrial exports and was particularly sensitive to maintaining good relations with the Empire (Rooth 1985). The conventional formulation was "the home producer first, then the Empire and finally foreign suppliers". But even when devising supports for the home producer, care was taken not to inflame Empire opinion – hence the device of government-subsidized deficiency payments rather than, as originally conceived, an import levy.

Following the outbreak of the Second World War, however, farmers were suddenly required to expand food production almost regardless of cost. With government now the sole purchaser of agricultural products, the price of cereals and milk was raised and subsidies were given for hill sheep. Farming prospered and this experience of a state-managed expansion of output proved a formative one for farmers and government alike. With the UK's unique dependence on imported food once more exposed by a world war, the seal was set on this new partnership.

Based as it was on polarizing and controlling the global pattern of the export of manufactured goods and the import of temperate and exotic foods for the industrialized workforce, the evolution of the Imperial food order had rendered the UK's rural areas economically marginal. The lengthy agricultural depression had led to thousands of acres of arable land lying unkempt and put under grass, and thousands of farm workers unemployed or suffering from low wages (Hall 1941). The National Farm Survey of 1941 revealed a graphic picture of the extent of the social and economic immiseration afflicting many farming families. Depressed and depopulated, many villages lacked the basic amenities that urban areas had long taken for granted. During its period of hegemonic imperialism, successive British governments had been prepared to cede the prosperity of the domestic agrarian population in their efforts to maintain international commercial preeminence and urban political support. The full consequences became ever more obvious as the non-agricultural sectors of the British economy lost their international competitiveness and dynamism, thus reducing off-farm employment opportunities for those leaving the land. With the whole of the Empire afflicted by the world depression, and the complete drying-up of the flow of investment capital from London, emigration to the Dominions and colonies, another traditional option for the rural unemployed, was also attenuated. Given the peculiar position of an unprotected British agriculture, the plight of rural areas and their people was tied inexorably to the turmoil in the international food commodity markets.

The Atlanticist food order, 1940–1970s

Recognition of the strategic importance of a strengthened domestic agriculture, in the wake of the U-boat campaign, ensured agreement between the political parties and all agricultural interests that there should be no return to the agricultural policies of the 1930s once war had ended. In any case, the acute shortage of food immediately after the war, and its lack of dollars, meant that the country was unable to purchase all its food requirements on world markets. As a result, the government continued to rely on British farmers to meet as much home demand as possible while imposing stringent food rationing to control demand. In order to save dollars and to reduce a massive balance of payments deficit, the Treasury called for a large increase in agricultural production and provided the extra finance to encourage it. As a Treasury official wrote in a letter to a Ministry of Agriculture official (Public Record Office, CAB 124/572, quoted in Smith 1989):

> the prospect of a dollar shortage has created the greatest opportunity for British agriculture that has occurred in a time of peace for a hundred years . . . [W]e are now in the position where agriculture will be under fire for not expanding enough . . . In these circumstances the time may come when certain advances which have hitherto been regarded as visionary may become practical politics.

The UK's postwar dollar crisis served to underline how financially (as well as militarily) dependent the country had become on the USA. Under US leadership a new global economic order was framed, including an international food order based on the US model of the relationship between industrial and agrarian development (Goodman & Redclift 1991). It was within this radically different context that the UK's postwar agricultural policy was developed.

During the interwar years, US opinion had seen the UK as a major trade rival. The Ottawa agreements on Imperial preference and the sterling area system had, in particular, caused great resentment among Congressmen, farmers and businessmen (Kolko 1968). So, when the Roosevelt administration began to look to establish more liberal trading conditions, the UK's trading arrangements were the most obvious target. Together, US and British imperial and sterling-area trade accounted for almost half of world trade. If the UK could be cajoled into agreement, other countries would be obliged to follow.

The British dependence on US productive capacity for the conduct of the war soon provided the US administration with a lever to press for the liberalization of the UK's postwar trade and payments. Through negotiations over lend-lease, then over postwar commercial and monetary policy, and finally over the terms of the UK's postwar dollar loan, the US Treasury took advantage of its growing economic power and the UK's increasing vulnerability to impose its particular vision of a liberal and expansionist international economic system (Dobson 1986).

The dollar loan was negotiated by the postwar Labour government after the abrupt termination of lend-lease in August 1945. From that moment on the UK had to pay for everything that the USA supplied, leaving it in what Hugh Dalton, the Chancellor of the Exchequer, described as "an almost desperate plight" (quoted in Harris 1982: 271). The country had been drained of its reserves during the war, was massively indebted and internationally over-committed, and faced major difficulties demilitarizing its economy. Labour, though, had swept to power on a programme of radical welfare reform. To be able to carry out its programme it had no alternative but to seek a loan from the USA.

However, the mood in Washington was hardening: there was considerable wariness about the new British government's commitment to planning, socialization and state trading; and questions of national advantage increasingly dominated US political debate while the loan was being negotiated (Booth 1990, Kolko & Kolko 1972). On the US side: "The loan was a means to prise open the sterling area and curb the Labour government's enthusiasm for intervention" (Booth 1990: 147). The negotiations were protracted and the British Cabinet several times considered breaking them off. Ministers became alarmed at the prospect of the repayment burden (Pimlott 1985: 430–1), close US supervision of UK policy (Dalton 1962: 80) and the clash between external and domestic policy goals. But they had little choice other than to accept the US terms. The loan, agreed in December 1945, was much smaller than had been hoped, was interest-bearing and came with conditions concerning sterling convertibility, sterling balances, import policy and the ratification of the Bretton Woods agreement. The external constraints thus imposed placed a damper on the Labour government's domestic policies.

One of Labour's most radical commitments was towards agriculture. Major legislation had been promised and ministers came under criticism from their own supporters as they delayed action through 1946. The party's position had been set out in *Our land*, published in 1943. This document called for the maintenance of wartime controls over farming, reinforced by large-scale land nationalization. Efficient, rational production was to be ensured by fixing the price of food "at the level necessary to bring forward the required national supply from farms run at a fair level of efficiency" (p. 8). Farm workers were also promised wages pegged to those of other skilled workers, and the abolition of the tied cottage.

But while the legislation was being prepared, international considerations increasingly preoccupied ministers. A world shortage of grain had produced a severe food crisis in 1946. The UK was obliged to supply not only its own population but also India and the British zone of occupation in Germany, where there was a threat of famine. To meet these additional commitments, the government renegotiated its allocation from the Washington-based Combined Food Board, which controlled the distribution of resources. Even so, it faced a reduction of 200,000 tons of wheat in the UK allocation for the summer months. Bread rationing, which had been avoided during the war, had to be introduced and, despite its unpopularity, lasted for two years (Flynn 1989). These difficulties were compounded by a looming economic crisis. In the circumstances, the US loan lasted only half as long as was intended, and by late 1946 it was becoming apparent that the impending requirement attached to the loan – to make sterling freely convertible into dollars – would be a perilous act. Nevertheless, ministers refused to make a formal request to delay convertibility, fearing that the last tranches of the loan might be withheld and that the anti-British and anti-socialist sections of the US Congress would use the opportunity to demand changes in British domestic policy (Dalton 1962: 254–6).

The food and financial crises focused attention on the need to boost domestic food supply as rapidly as possible. As the UK imported so much food, the expansion of domestic agriculture was potentially a prime dollar saver, and this became the overriding policy preoccupation. But after years of depression and wartime demands, the agricultural industry was desperately short of capital for equipment and livestock, and as the Labour Minister for Agriculture recorded (Williams 1965: 164):

It seemed close to impossible that any Government would be prepared to spend the necessary extra sums to inject this capital . . . However, the Cabinet took the right decision and a bold one. They agreed to add £40,000,000 *each year* for the next four years to the guaranteed prices negotiated at the February price review to provide capital for an agricultural expansion programme.

This commitment to agricultural expansion was indeed extraordinary for a country entering "the age of austerity", but external financial circumstances dictated it. Perhaps even more extraordinary was that it came shorn of socialist trappings. As part of a more general shift after 1946, which saw the tempering of ministers' redistributive aims and moves towards greater pragmatism, the 1947 Agriculture Act dropped most of Labour's more radical commitments. It normalized the war-time system of guaranteed minimum prices and assured markets for certain staple products, and formally committed government to consult farming leaders in an annual price review. But measures to nationalize land, direct production, abolish tied cottages or link farm workers' wages to a general wages index, were abandoned. These policies might alienate farmers or, for that matter the US administration, both of whose cooperation was now vital to national recovery. In this respect, the consequence of the UK's dependence on imported food was to expose its vulnerability, not to its wartime enemies, but to its peacetime ally.

It fell to the Conservative government elected in 1951 to lift many of the wartime controls, but the policy of agricultural expansion and support continued. From 1953, agricultural imports were allowed into the country without duty and farmers were compensated for the difference between market prices and annual guaranteed prices through a system of deficiency payments. Cheap imports thus helped to keep food prices low, and former colonies happy, and the support was funded indirectly by general taxation. However, this approach placed a very considerable burden on the Exchequer, a burden that grew relentlessly as farmers increased their productivity and international food prices fell as world farming recovered from the disruptions of war. As postwar food shortages receded, therefore, the emphasis of policy switched from simply raising output to encouraging increased efficiency, particularly of labour, and containing the costs of price support.

The drive to efficiency was two-pronged. First, the government directly financed agricultural research and education, and a state advisory service. Particular emphasis was placed on the development and promotion of labour-saving and yield-increasing technologies. Secondly, support policy was oriented towards encouraging farmers to adopt the new technologies. This was done directly through capital grants and input subsidies, and indirectly through a steady squeeze on guaranteed prices, though in 1956 the Conservative government agreed not to reduce the total value of support prices by more than 2.5 per cent a year. Even so, from the late 1950s the proportion of government expenditure on producer subsidies increased steadily and that on guaranteeing prices diminished, despite opposition from the farming lobby. For many farmers, this changing balance of support acted as a powerful incentive not only to adopt new techniques to boost production, but also to cut costs through releasing labour or enlarging their holdings, and to borrow the money to finance land purchase and capital projects. It is no coincidence that average farm holding size remained unaltered between the 1880s and the early 1950s but increased rapidly over the ensuing three decades (Grigg 1987).

Parallel to these steps to improve the output, efficiency and security of domestic agriculture, a concerted effort was made to expand world food trade in the immediate postwar period. The USA took the lead, as the largest food exporter and as part of its expanding economic hegemony, although the UK, as the largest food importer and the former orchestrator of global food markets, also played a significant rôle. Together, the two countries effectively blocked ambitious proposals brought forward by the first director-general of the Food & Agriculture Organization (FAO) for a World Food Board to plan and regulate global food supplies (Peterson 1990). They ensured that the integration of food trade would occur, if at all, through market liberalization and the agencies of investment, trade and aid, all of which favoured the USA with its economic power and ability to use the dollar's rôle as an aid mechanism.

Some commentators have talked of the establishment in the postwar period of an "international food order" under US leadership, which brought a remarkable period of stability to world agricultural markets in the 1950s and 1960s (Goodman & Redclift 1989, Friedmann & McMichael 1989, Kenney et al. 1989). As Friedmann & McMichael point out, the USA was pre-eminent among a number of settler societies that became central to the world economy in the 20th century. They had all been integrated into the global economy through export-oriented agrarian development that had then provided the basis for the development of domestic industries (McMichael 1984). Indeed, in contrast to the separation of industry and agriculture in the British model during the Imperial period, the economic development of the settler states had been based upon a reciprocal relationship between agrarian and industrial development and thus a national model of accumulation with "articulated" sectors. This relationship was well expressed in the early development of industrialized forms of agriculture, dependent upon mechanical and chemical inputs. More specifically, by the 1940s, the US administration was expounding a model of technological innovation and market intervention for agriculture to be disseminated internationally. The technological base for this capital- and energy-intensive model of agro-industrial accumulation depended on a flow of genetic innovations, beginning with the hybridization of corn in the 1930s. The convergence of mechanical and agrochemical developments on plant genetics gave rise to complementary, integrated technological "packages" in major crop sectors (Goodman et al. 1987).

The New Deal, with its programmes of protection, price stabilization, farm-income support and investment incentives, provided the initial context for the regeneration of US agriculture and its technological transformation (Gilbert & Howe 1991). Subsequently, under the Marshall Plan, a framework was established for the reinvigoration of western European agriculture as well as industry, which depended on the stimulus of the market as well as the political backing of the state. The US Treasury and the State Department insisted that aid to Europe should be conditional on the removal of internal barriers to trade and the movement of factors of production (Milward 1984). In addition, an evolving structure of price supports and production incentives established the foundations for sustained growth of output and productivity. The benefits were widespread, including a general revival of rural prosperity based on agricultural production, a considerable stimulus to the expansion of the agro-industrial sector and off-farm sectors of the food and fibre system, and the development of mass consumption policies centred on basic food and feed grains and livestock products. The model of agricultural regulation and technological competition that emerged from the New Deal and wartime emergency controls thus played an important rôle in the postwar consolidation of the Fordist regime of accumulation (Kenney et al. 1989). At the same time, the internationalization of production and accumulation in the world economy led to the integration and interdependence of food systems, with increasing interpenetration of national markets, technological convergence and more uniform patterns of food consumption. All these developments have been promoted by international agribusiness, whose interests cross the whole globe, spanning political systems and core and peripheral economies.

The productivist agricultural regime in the UK

In the postwar period, the UK also began to emulate the model of a national economy with its notions of articulated sectors and balanced growth, and nowhere was this more apparent than in the promotion of the agricultural sector. The system of deficiency payments, the openness of the British market to food imports, and the continuation of strong trade links with the Commonwealth were redolent of the UK's former Imperial rôle. But the expansion of domestic agriculture was also relentlessly pursued and its industrialization became a focus for accumulation. This involved a new coalition between domestic farming and industrial interests as part of a productivist agricultural regime.

The productivist regime was forged in the 1940s. It co-existed with, but predominated over, a social welfarist regime of rural support that had its roots in the 1930s. The latter was pursued mainly in the remoter rural areas of the north and west where the climate and terrain severely limited the scope for agricultural intensification. Unable to contribute a great deal to urban food needs, many of these wild marginal upland areas were allocated secondary functions of forestry development, water catchment, informal recreation and nature conservation. Assistance to farmers in these more remote areas reflected farmers' needs for support (i.e. a welfare policy). For example, the 1946 Hill Farming Act continued the payment of hill livestock subsidies introduced during the Second World War. This type of payment became an enduring feature of peacetime policy, eventually being encapsulated in the European Community's (EC) Less Favoured Area Directive of 1975, but the tensions between welfare support and improved farming efficiency are reflected in the contradictions contained within other legislation. The 1957 Agriculture Act established a Farm Improvement Scheme, targeted on hill and upland farms, which gave grants for land improvement and reclamation, while the Small Farmers Scheme of 1958 encouraged the adoption of farm business plans. Both schemes aimed to support the hill farmer and small producer, but only through the search for greater business viability, a process that many upland farming households were unable to sustain. A smaller number of larger upland farms have, therefore, been the greatest beneficiaries of a policy justified primarily on welfare grounds.

Nevertheless, the scale of state support for these areas led to the formation of close industry–government links (Murdoch 1992) which we have termed clientelism (Ch. 8). On the other hand, several key elements acted to secure the productivist regime:

○ *Security of land rights* A productivist agriculture needing fixity of productive capital required security of tenure. The Agricultural Holdings Act 1948 ensured lifetime security for tenants. In political terms, this placed the seal on the ascendancy of productive capital (represented by the National Farmers' Union – NFU) over landed capital (represented by the Country Landowners' Association – CLA), but it contained the seeds of its own contradictions. By reducing the number of new tenancies, it created an insider group' of *in situ* tenants and, indirectly, elevated freehold ownership into an objective in its own right (Ch. 4). The high-water mark of security of tenure was the 1976 legislation which, under certain quite liberal conditions, introduced statutory successional rights for three generations of tenants from within the same farming households. Other rights were even more specifically a function of state support in that they regulated access to that support. These included, for example, the official status of a farm as a registered agricultural holding, a part-time or full-time holding, a hill farm, a dairy farm, and so forth.

○ *Security of land use* The 1947 Town and Country Planning Act established a secure physical environment for agriculture by according to farming a pre-emptive claim over rural land and, by excluding other employers, a relatively cheap source of labour. Not only did this place constraints on the pace and location of development affecting rural land, but also helped establish a rural planning hierarchy with the Ministry of Agriculture, Fisheries and Food (MAFF) at the pinnacle. From its dominant position, it watched over a series of subsidiary agencies, including local planning authorities, water authorities, the Nature Conservancy, national park authorities and the Forestry Commission, to ensure that in their land-use planning functions the needs of agriculture were safeguarded (see Ch. 5 for a fuller analysis of these issues).

○ *Financial security* The 1947 Agriculture Act formally established a system of annually negotiated guaranteed prices for almost all staple products. On the one hand, this offered some protection to farmers from market fluctuations and cheap imports while encouraging them to produce more. On the other, it enabled the state both to manage the expansion of the agricultural sector and to maintain low prices for the consumer through a system of deficiency payments. The system preserved a delicate balance between the state's overall management of the sector and the autonomy of the individual producer. However, the financial security offered was for accumulation within the sector (rather than for individual producers) and in the context of an unregulated land market it facilitated the steady concentration of production.

○ *Political security* The key position of the NFU in postwar agricultural policy must be seen as a function of its political significance rather than of economic power. In other words, its influence has stemmed not from the market strength of its members – a multitude of small producers – but rather from effective organization in the context of a politically prescribed partnership between government and the farming community. In this partnership, the NFU was accorded a key mediating rôle: representing the interests of the farming community in policy-making and helping to ensure the support of the farming community in the implementation of agreed policies. Corporatist arrangements such as these, involving exclusive political access and privileged receipt of government largesse, needed to be bolstered by a legitimating ideology.

○ *Ideological security* This ideology has been furnished principally by the NFU and MAFF, along with the agricultural research and advisory establishment, the farming press, and less visibly, but no less importantly, by industrial capitals operating in the agricultural inputs and food-processing sectors of the food system. It has included a number of separate strands. The first is identification with the national interest, which over the years has encompassed arguments ranging from food security to the balance of payments. The second is a productivist image – that farming is an expansive, efficient, modern and technologically advanced industry (Wormell 1978). The third is a voluntaristic and individualistic image of farmers as competitive and independent-minded entrepreneurs. Aside from these dominant strands, there have been other subordinate justifications for postwar agricultural support used in more limited contexts. Farming, for example, has been presented as an activity that was essentially protective of the rural scene, and one which was the source of prosperity for rural areas and for rural people. These basic tenets of the productivist period were maintained largely intact until at least the late 1970s. They still dominated the thinking contained in the White Paper *Farming and the nation* (MAFF 1979), even if less dogmatically than the earlier White Paper *Food from our own resources* (MAFF 1975). These policy statements underlined the stability of rural land regulation during this period. While rural planning was designed both to protect rural land and to confer upon its owners the freedom to farm it as they wished, agricultural policy was to provide the wherewithal to ensure accumulation and growth in the intensity of production. As we shall discuss in Chapter 4, the rural landed interest played a key rôle in the development and maintenance of this policy. Indeed, rural landowners, and particularly the increasing number of owner–occupiers, were to be its principal beneficiaries. There is little need here to outline many of the consequences of this policy for British rural society, given its considerable treatment in the literature. What is significant, however, for our analysis of contemporary rural change is the basic causes of its demise and its reconstitution during the 1980s.

The onset of contradictions: change in the 1980s

Fissures in the Atlanticist food order became apparent from the early 1970s. Just as industrial capital generally was to experience a massive upheaval in the face of the oil price rises of that decade, the food sector increasingly became an arena of uncertainty and struggle between nation states, both North–South and East–West. Trade wars became more frequent, and the tensions between national economic and political strategies on the one hand, and between multi-national corporations on the other, introduced increased instability into food commodity prices (Goodman & Redclift 1991). Coincidentally, rising concern for food quality and the environment, as well as a growing critique of agricultural policy from ascendant New Right politicians, alarmed at the cost and seeming inefficiency of public expenditure on agriculture, began to undermine the state-sustained productivist model. In spite of rising expenditure on the EC's Common Agricultural Policy (CAP), most farmers' incomes fell in real terms, even though politicians and taxpayers presumed farmers to be the primary beneficiaries of the policy. Once land prices also began to fall, as they did in the UK from 1980, then the major gain accruing to owner–occupiers since the mid-1950s from the state support of farm commodity prices also began to fade.

These economic weaknesses emerged just as the political power and security embodied within the postwar partnership between the NFU and MAFF became increasingly strained; indeed the two processes were interrelated. Internal conflicts, based on interests represented by different commodities and scales of production, became more frequent in the business of the NFU; and its overall rôle in influencing prices through the review procedure began to wane, exacerbated by the shift of some decision making from Whitehall to EC headquarters in Brussels. The EC's sudden introduction in 1984 of milk quotas to curb overproduction embarrassingly exposed the impotence of both MAFF and the NFU in the face of new political priorities and loci of decisions (Cox et al. 1987).

Finally, the credibility of the dominant ideology, which had allowed farmers to remain protected (through planning policy) and supported (through agricultural policy) but to avow a voluntaristic and independent ethos in terms of environmental and rural policy, became increasingly strained. Critics characterized such a set of circumstances as "theft" (Shoard 1980) or as "shameful" (Body 1982). The political management of these previously buried contradictions came to preoccupy policymakers during the 1980s. Agricultural land use, oriented towards increased yields, had been swathed in consensus, but it now became a sharply contested issue (Bowers & Cheshire 1983).

Of course, the productivist regime had been beset by contradictions long before the 1980s. The art of political leadership of the agricultural community in the postwar period had been to manage the steadily mounting tensions inherent in a policy at once protectionist and expansionist. What was new in the farm crisis of the 1980s was the overwhelming build-up of contradictions and the way these spilt over from domestic and EC politics into international politics. The globalization of the food crisis emerged with the collapse of the international food order that had been established under the USA's former hegemony of world trade, and the increasingly cut-throat competition that emerged as countries sought to offload their growing agricultural surpluses on overseas markets in a mounting trade war. The consequent depression of world commodity prices drove up the costs of disposing of surpluses and fuelled the fiscal crisis in agricultural support, a crisis now accompanied by various conjunctural features such as farm bankruptcies and depressed land values. But, at its roots are the difficulties occasioned by an expansion of agricultural output at a rate that has outstripped the capacity of domestic markets (as well as the capacity of natural systems) to absorb the increase, meaning that an alternative social justification must now be found. Both the state and agricultural interest groups accept the necessity of reducing the level of overproduction and containing public expenditure, but the political difficulties of achieving this in the face of entrenched farming interests across EC member states is all too evident (see below).

As Goodman & Redclift (1989: 12) remark: "The farm crisis has exposed current production regimes to a crisis of legitimacy". Indeed, because agriculture has for so long been an extensively supported industry, the present crisis reveals itself in the form of a "crisis of crisis management", to use Claus Offe's term (1984). Crises arise from unresolved steering problems and their defining feature, as both Offe (1984) and Habermas (1976) have argued, follows from the way in which actual or potential change threatens social identity. The farm crisis is therefore simultaneously political and ideological as well as economic (Pile 1990, Cox et al. 1989).

In order to make sense of the farm crisis in the British context it is necessary to understand the contradictions within the dominant Thatcherite political philosophy of the 1980s and between it and the productivist ideology of the agricultural industry. The profound antipathy of the New Right towards corporatist politics has meant, most notably, the exclusion of producer interests and, especially, organized labour, from a strategic rôle in macroeconomic policy making. The resilience of corporatism at the meso-, or sectoral, level should not be underestimated, however. Some of such arrangements have survived and, in certain cases, even flourished, reflecting in part the strength and continuity of government–industry links in particular sectors (Cawson 1986).

Some aspects of Thatcherite reformism have actually served to promote meso-corporatism at the sectoral level since the emphasis on "rolling back the frontiers of the state" and asserting the primacy of the market have led inadvertently to the development of self-regulatory mechanisms embodying corporatist relationships. Thus, the introduction of milk quotas in April 1984 – only agreed to by the government to relieve the budgetary crisis of the

CAP – served, for a while at least, to strengthen the rôle of the Milk Marketing Board, itself a distinctly corporatist agency. In the sphere of land-use politics, moreover, the promotion of voluntary co-operation and self-regulation, as the solution to the problems of the relationship between agriculture and conservation has further extended corporatist relations (Cox et al. 1990). Indeed, the institutional forms, networks and norms that constituted the corporatist character of the productivist era have persisted, even as the ideology with which they were associated has been increasingly compromised.

Thus, although at the level of rhetoric it is evidently the case that New Right pronouncements were often deeply uncongenial to the agricultural industry, other contradictions within contemporary Conservatism limited the scope for the sort of onslaught on agriculture that was visited upon other traditional industries in which the state had become deeply involved, though even with them there was often a gap between rhetoric and achievement (Gamble 1988, Jessop et al. 1989). Agriculture and the countryside have in any case long occupied a special place in the pantheon of traditional Conservative values and, at the level of symbolism certainly, rural areas remain emblematic of the Conservative heartland. The impulse to relax planning controls and to force farmers increasingly to accept the dictates of the market in order to survive is apparent, but it finds little favour among middle-class Conservative voters who have moved to the countryside and are keen on rural preservation. In other ways, too, shire paternalists find themselves radically at odds with right-wing free-marketeers. Ideological and electoral considerations have therefore placed bounds on the impact of neoliberal radicalism on agriculture and, as we shall see in Chapter 5, on rural planning.

In addition, the bureaucratic and complex EC structures for agricultural policy making, in the form of the CAP, greatly constrain member states' autonomy. The UK is locked into a strongly corporatist approach to decision-making on farming issues, as well as one that embodies a social welfare commitment to farming. So although extensive farm support remains a matter of considerable embarrassment to the government, and is repeatedly attacked by right-wing backbenchers and think-tanks for its bearing on food surpluses, public expenditure levels and the liberalization of world trade, the scope for manoeuvre is seldom more than limited. The Thatcherite offensive has, in short, had the effect of diminishing the legitimacy of agricultural support in the UK without seriously compromising established meso-corporatist arrangements (Cox et al. 1990).

The notable contradictions within contemporary Conservatism are matched by fissures within the dominant agricultural ideology. The recent sustained attention directed at the excesses of agricultural policy has greatly undermined the plausibility of its productivist ideology. The farming and landowning lobby has reacted by working to embrace alternative ideologies and to demonstrate some commitment to rural welfare and conservation. There are, after all, other resources available to be mobilized. The landowning community might plausibly claim, for instance, that its commitment to stewardship has only been possible because of a tradition of imaginative asset management and diversified activity in the past. Such adjustments are not easy, however, and the effort is handicapped by the legacy of the productivist ideology which is deeply ingrained in the outlook and behaviour of many farmers, landowners and agricultural officials. There are, in any case, limits to the speed with which alternative characterizations can be readily embraced, especially as there are no price-support mechanisms associated with many of the activities that might nurture a post-productivist image.

These transitional circumstances have presented the British government with difficulties in positioning itself in relation to negotiations over the future of state support for agriculture at the global and EC levels. It strongly endorses attempts to reduce public expenditure and to make production more closely related to demand, but at the same time does not wish to be seen to undermine the interests of its own farming community.

On the global stage, the round of GATT trade negotiations, which started in 1986 and were due to end in 1990, became deadlocked over the same issue of agricultural trade. The protracted negotiations not only exposed the extensive involvement by governments worldwide in the management of their agricultural sectors but also the continued decline of US dominance of the world economy. The USA was a strong supporter of the incorporation of agriculture in the Uruguay Round, allying itself with the Cairns Group of countries heavily dependent upon agricultural exports (including Australia, Brazil, Canada, Chile, Colombia, Hungary, Indonesia, Malaysia, New Zealand, the Philippines, Thailand and Uruguay). Together, they have challenged the agricultural protectionism of the EC and Japan. Throughout the protracted negotiations, Japan and the EC have resisted the radical proposals of the USA and the Cairns Group which entail the almost complete dismantling of current measures (internal support, import access and export subsidies) that distort international trade. Considerable market deregulation has already been effected unilaterally in Australia and New Zealand (Le Heron 1991, Lawrence 1990). In reply, the EC has proposed modest reductions to the current level of support (e.g. by 30 per cent over 10 years), seeking to deny the "level playing field" sought by a USA confident that its large, technologically advanced farmers would be very competitive in international markets (Byman 1990). The British government is more sympathetic towards the US position than most of its EC partners, since it believes that its farming sector can compete, but wants to phase in price reductions to allow its farmers time to adapt in a measured and responsible way. But it finds itself on the opposite side of the table from many of the countries that first developed agricultural exports to supply the British market.

Within the EC, broad recognition among all member states that levels of production must be brought more into balance with market demand is confused by each of them playing for national stakes. The end result is that the current proposals for reform of the CAP (the "MacSharry Proposals" – European Commission 1991) seek to marry social and environmental objectives with reduced market protection. In its original proposals, the European Commission sought to taper farm support in favour of small producers on grounds of equity, maintenance of a "rural society" and environmental protection. But in the agreement first reached (May 1992), and for which details are not yet available, this principle has been very largely discarded, much to the relief of British farming interests, which would have been most severely disadvantaged because of the large average size of British farms (Allanson 1992).

As important, there will, *inter alia*, be reductions in support prices for key commodities (e.g. cereals, 29 per cent; milk, 10 per cent; beef, 15 per cent), and milk quotas will experience another small reduction. In the short term, at least, farmers are to be fully compensated for these reductions through full payment for any resources transferred from farming (by an extended set-aside scheme for cereals, additional afforestation, and greater early retirement incentives for farmers, for example), thus *raising* rather than lowering the cost of the CAP. The British government endorses aspects of the package, such as the reduction in price-support levels and additional funding for environmental protection, but is very concerned at the rise in total cost. For example, prior to agreement, it was estimated that expenditure under the existing FEOGA Guarantee Section would rise by 20 per cent in 1991 and a further 11 per cent in 1992 to Ecu 36 billion (MAFF 1975). Moreover, the compensating measures are open to serious fraud, as the payments will be largely dependent on information supplied by farmers and the means of checking claims are rudimentary.

The conflicts between EC member states and between the interests of producers, consumers and environmental groups, have led farming leaders to seek other means of refurbishing the ideology of agricultural support. The most sustained attempt has been via the debate on alternative land use and promotion of farm diversification. Since 1984, the terms upon which the farm crisis has been debated have shifted from food surpluses to a prospective land surplus. This shift has been premised on the assumption that if production controls, such as milk quotas, proved effective and were introduced for other staple commodities, some agricultural land would no longer be needed in production. The shift first occurred in agricultural circles, then in the farming press, and by 1986–7 it was reflected in the national press. Increasingly, policy options in response to the farm crisis were posed in terms of the need to promote farm diversification and alternative land uses (Cox et al. 1988).

Although most of the work on alternative crops and animal products was conducted by agricultural and environmental scientists (e.g. Carruthers 1986), and on the future agricultural land budget by agricultural economists (North 1988, Potter 1991), the debate has had distinct political and ideological determinants. The farm crisis and its solution could have been diagnosed quite differently, and to construe it in terms of the search for and encouragement of alternative land uses was to make explicit or implicit choices that favoured certain interests (Lowe & Winter 1987). Economic analysis, for example, has tended to see over-production in agriculture as caused not by a surfeit of land in production, but rather factors such as chemical inputs or machinery, or capital in general (Body 1982, Bowers & Cheshire 1983). To have tackled these oversupply problems, through either market or bureaucratic mechanisms, would have depressed landed capital and agribusiness interests. Conversely, to incline policy choices and financial support to the redeployment of supposedly surplus land favours just these interests. Thus recent policy initiatives and proposals have included grant schemes to make cereal land idle, reduce the stocking densities of beef herds, plant farm woodland and support traditional farming practices in environmentally sensitive areas. There have also been efforts to relax planning constraints on the re-use of agricultural land and buildings.

The alternative land-use debate, by shifting attention from food surpluses and their fundamental cause to questions of farm diversification and novel uses for rural land, has helped to revive a productivist and innovative image for the agricultural industry, at least temporarily. Moreover, with the spectre of large-scale and unmanaged land abandonment deployed as a warning against an uncared-for and derelict countryside, environmental policy for agriculture has come to be orchestrated in terms of the same agenda, re-emphasising once again the stewardship rôle of rural landowners. In the process, pressures for fundamental reform of the agricultural support system have been largely dissipated. Indeed, the manner in which the alternative land-use debate has been conducted is itself an indication of the persistent power of constraint enjoyed by farming and landowning interests. Inevitably, though, the putative surplus of millions of hectares of agricultural land has opened a wider debate on the access of non-agricultural interests to rural land, thereby allowing a wider range of interests to stake claims, and further compromising the productivist ideology which is now so obviously in disarray (Countryside Review Panel 1987). How such claims are mediated is critical for the rural land development process since the access of non-agricultural capital to rural land is regulated by the planning system. But before it is possible to turn to these issues it is necessary to account for the long-term shifts in the pattern of rural property rights and the switch from agrarian landlordism to owner-occupation.

CHAPTER 4

Property rights
and interests in land

Introduction

In the previous chapter we outlined the distinctive approach
adopted by the British government towards agricultural regulation since the mid-19th century and how, through the UK's
changing position in the global political economy, this had led to
a rural development policy which, in broad terms, prioritized
national food security. This primary concern is reflected not only
in the evolution of agricultural policy but also in the objectives of
the postwar town and country planning system which sought,
inter alia, to contain urban growth and protect the agricultural
land base (Ch. 5). Neither agricultural policy nor the planning
system have paid much attention to the welfare of rural communities as such, in contrast to the objectives of rural policy over
much of Continental Europe. Rural policy goals have been
sought instead through the regulation of land use rather than
social provision for those groups that have traditionally earned
their living in the countryside.

One consequence has been a continuing debate in the UK over
the distribution of property rights. In spite of a long-established
land market, the ownership of rural land still confers social status
as well as economic power, making the allocation of property
rights a strongly contested issue. Conflict is further heightened
by high population density, considerable personal mobility and
the concentration of ownership in a relatively small number of
hands. In national economic terms, the ownership of rural land
may be of modest significance but the local distribution of
property rights remains crucial to the pattern and processes of
rural development. This is pre-eminently because of the continued association between control over property rights, local
elites and the rural class structure, and the focus upon land as
the means of realizing many public policy objectives. The clash
of interests is made especially visible by the functioning of the

statutory land-use planning system and related fiscal measures (as in the taxation of betterment or in the identification of planning gain) in the implementation of agricultural policy (in which subsidies have been paid to farmers on the basis of the amount of food they produce and not in relation to need), and in the execution of environmental policy (where compensation has been paid to owners and occupiers in order to restrain polluting or other destructive practices in proportion to the income from production forgone, instead of penalties imposed in accordance with the damage caused).

These contemporary conflicts have historical antecedents. The account that follows focuses upon the changing pattern of rural property rights since the mid-19th century and the ways in which different groups have sought to represent their positions. As in Chapter 3, the discussion is divided into two periods broadly representing the Imperial food order (1860–1940), and the more recent productivist phase in British agriculture (1940–85). We do not wish to suggest, however, a simple causal relation between change in the pattern and purpose of rural landowner-ship and the two periods identified. To do so would diminish the importance of other social changes. For example, the shift from a landlord–tenant system of agricultural land tenure to one of owner-occupation is but part of a more general shift in property relations in which legislation on tenure and taxation have, until very recently, worked against the interests of landed capital of all kinds. There is, moreover, a considerable lag in changes to property relations. One of the key characteristics of landowners throughout British history has been their ability to defend and then to adapt their interests in response to changing economic and social circumstances. This means that at the local level at least they have often been able to maintain a not inconsiderable presence, even if changes in the manner in which property rights are held and their extent has been profound. There remains a surprising degree of continuity in family ownership, the relative-ly closed landowning oligarchy of the 1890s having been replaced by a larger but still powerful group, consisting mainly of owner-occupying farmers (including many former landlords) in the 1990s.

In this chapter we attempt to do more than recount the changing pattern of landownership. We seek to demonstrate the differing interests and changing relations between landed and finance capital, and between land as a repository of cultural values and its use in production and exchange. In terms of land tenure, the increasing commoditization of land is reflected in the shift from the objectives that underlay a former landlord–tenant system to one dominated by owner-occupation, and more recently to signs of an ever greater subdivision between interests (family and non-family) of the bundle of rights held under the freehold ownership of a particular property (Ch. 2). Furthermore, the need for flexibility and divisibility of rights, in order to allow a speedy response to new economic opportunities, is viewed as part of a larger compromise constantly being struck between existing and new interests. This process has allowed for greater continuity of ownership than might otherwise have been expected, but achieved only at the expense of a reduction in private control, the result of complex management agreements between owners and finance capital and environmental interests. Leading to, and arising from, these altered distributions of property rights are altered modes of representation among those who wish to claim a stake in the future of the countryside. In this discussion, attention is focused on central government as the regulator of property rights (e.g. land tenure, taxation and environmental legislation) and the attempts of private property-owning interests to mould the pattern of regulation to meet their own objectives.

Rural landlordism in retreat, 1860–1940

The single most important feature of rural landownership in the past 150 years has been the continuous decline in the economic and political power of rural landlords. Their relationship to land was as rentiers. Only some were actively involved in the day-to-day management of their estates and even fewer were personally engaged in farming. Most of their land was let to tenant farmers. In the countryside, their economic power derived from control over access to land and a legal system weighted in their favour, while their political power stemmed from dominant representation in Parliament and in the major offices of state, and control over county government. Many had economic interests beyond their rural estates, in investments at home or abroad, including urban property. These resources often provided the means to sustain their rural estates long after agriculture had gone into recession in the late 1870s. The ownership of rural estates continued to provide status, acted as a symbol of continuity and, on occasion, reinforced a direct link with the aristocracy. Associated with these non-material trappings of social authority was the notion of "stewardship". In describing the notion Newby et al. (1978: 23) draw on the ideas of Edmund Burke who opposed the mere principle of utility, arguing that "in the context of the historical 'process of nature' man's transitory and fleeting existence reduced him to the status of a steward serving and caring for the landed estate which transcended the generations . . . The 'steward' therefore served rather than owned the property".

In more mundane terms, stewardship might be taken to mean the prudent development and management of an estate with an eye to family continuity, a concern to avoid saddling it with debt, its embellishment through the construction of grand houses and associated landscapes, and a system of local patronage through which owners might claim to protect the interests of those living and working on their estates. The actual behaviour of owners reveals a more pragmatic outlook. To sustain or build up a large estate often involved careful alliances between landed families that were usually sealed through strategic marriages. To profit fully from one's estate, indeed to support a style of living in accord with a landowner's rank, also demanded a flexibility towards estate management and improvement placed a premium on the services of astute and progressive land agents. From the late 19th century, however, concentrated landownership faced increasingly hostile political and economic conditions; survival in such a hostile world certainly demanded adaptability and business acumen.

Indeed, despite the radical changes to landownership that have occurred during the 20th century, many landowning families have been able to sustain certain of their interests. Their retreat from a position of pre-eminence has often been marked by an astute manipulation of the dwindling economic and political resources available to them, seizing what opportunities came their way. The divisibility of property rights has permitted flexibility of management in the face of a widening range of demands made by the state and financial and industrial capital, as well as the changing objectives of landowners themselves. It is also apparent in the tactics of their political representatives: knowing how and when to contest the retreat.

At the end of the 18th century, the landowning and aristocratic elite occupied an extremely powerful and entrenched position in the economic and political life of the nation, protected from dilution and break-up from within by the widespread practices of primogeniture and the settled estate. In spite of this, the 19th century saw a dramatic weakening of this elite's economic power because of the transfer of economic activity from the countryside to the town. The effects were variable: many of the more financially alert invested in the industrial and commercial sectors of the economy at home and overseas, or were able to profit from urban growth or the exploitation of mineral rights.

Moreover, the decline in the social authority and political power of the rural landlord lagged significantly behind economic changes. Even in the mid-19th century, two-thirds of the members of the House of Commons owned rural land (Sutherland 1988), and landowners remained in a majority in the House of Commons until 1885 and in the Cabinet until 1906. But behind the scenes their power was being undermined: the Reform Acts of 1867 and 1884 widened suffrage, and in 1872 the introduction of the secret ballot reduced the power of patronage. Together, these changes helped fuel radical opposition to landowning interests that over the ensuing half century was able to alter the public perception of the landed elite from one of deference to an awareness that what was regarded as excessive and unearned wealth might be taxed for the benefit of the nation, or even be replaced wholly or in part by a countryside given over to small holdings (Offer 1981). The establishment of school boards and sanitary boards in 1870, and county councils in 1888, also altered landowners' relations with their local communities. Individually, they might be appointed to key positions, but their power was increasingly mediated through the democratic process and by a growing professional class of local officials (Lee 1963, Mingay 1990).

The impact of these changes on the structure of landownership was slow to materialize. The 1873 *Return of owners of land* revealed that fewer than 7,000 of them held the freehold rights to more than 80 per cent of the nation's land, with the largest 1,500 owning 43 per cent of the total (Bateman 1883). There is little reason to suppose that this picture changed markedly during the rest of the century. In 1900, the landlord–tenant system still dominated the pattern of agricultural land tenure, with nearly 90 per cent of land let on this basis.

The fragility of this dominance was brutally exposed by the prolonged agricultural recession that began in the late 1870s. It undermined the economic position of the landed estate and created the conditions for population decline across much of the countryside. The prosperity of the period 1850–75 had encouraged landowners to borrow and by "1875 between two-thirds and three-quarters of the long term debts secured on landed estates were probably owed to insurance companies". These companies did not need to "invest directly in agricultural land at that time, largely because, despite rental yields of around 2.5 per cent, private owners were prepared to pay 4–5 per cent on borrowed capital" for the privilege of ownership (Northfield Committee 1979: 29). The long-established tradition of paying a high price for investing in land had made those recent entrants, who were largely dependent for their incomes on their landed estates, very vulnerable to the gathering recession and to the interests of banking capital. Furthermore, although not of great economic consequence at the time, the disputes over tenants' rights in the 1880s weakened the social control of the landlord while strengthening the economic position of the tenantry. By providing compensation for a tenant's improvements on the termination of a lease, a key element in the security necessary for the development of industrialized agriculture was put in place.

In a pragmatic response to these pressures, and in a move that aided the survival of the financially adept, the Settled Land Act (1882) was passed. This allowed the "life-tenant" (i.e. owner) of a settled, landed estate to engage freely in the sale, lease and transfer of land that had been inherited, as well as the land market more generally. By the turn of the century, only between one-quarter and one-third of estates by value were settled (Offer 1981), compared to an estimate of approximately 50 per cent made by Thompson (1963) for the mid-19th century. The change provided some necessary flexibility for the landowner, but at the same time weakened the hold of established landowning families over rural land. The *nouveaux riches* could gain access to land much more easily through the marketplace and this opportunity contributed to the final, glorious fling of the landed estate as many successful entrepreneurs of the late Victorian and Edwardian periods sought respectability through land purchase and enjoyment through country pursuits. This process was especially evident in the Highlands of Scotland, the new purchasers sustaining and reinforcing a local pattern of ownership that was the most highly concentrated in the UK. In 1884, 104 estates covered 2 million acres, while by 1912 there were 203 large estates occupying 3¼ million acres largely devoted to shooting (Bryden & Houston 1978, Smith 1992).

The distinctiveness and potential transience of this revised social order is reflected in Clemenson's description (1982: 104) of the period:

landowners not dependent upon agricultural income, sublimely unaffected by the impact of the depression, maintained the opulence and landed traditions of the high-Victorian era. The pleasures of hunting and shooting, of seasons in London and lavish entertainment at the country home devolved to the *nouveaux riches* and the more wealthy established landowners.

This brief revitalization of the landed estate, dependent on industrial and banking capital and overtly based on consumption rather than production objectives, provided tangible expression of a changed rural world to which the urban bourgeoisie looked for retreat and cultured enjoyment (Wiener 1981). The middle classes generally were increasingly appalled by 19th-century urban conditions. Explorations into the conditions of the poor in the cities were unambiguously pessimistic (Rowntree 1915). The countryside, by contrast, seemed to represent everything the city was not: it offered rustic peace and tranquillity, an escape from the "dirty utilitarian logic of industry and commerce" (Chambers 1990: 33); and no sooner did enquiries suggest that all was not well with rural living conditions (Land Enquiry Committee 1913) than established interests leapt to their defence (Land Agents' Society 1916). As the economic base of the countryside declined so its cultural significance grew, a process that has extended into this century.

It is against this background that we can trace the rise of "preservationism" as a social and political movement in the post-Darwinian world of the late 19th century. Nature came to be seen as increasingly vulnerable in the face of population growth and industrial capitalism, creating an ecological consciousness that appeared alongside an urban middle-class concern to protect traditional rural landscapes. At the same time, working-class inhabitants sought to make use of, or to establish, customary rights of access over open countryside around industrial towns and cities to provide a temporary escape from their dreary and polluted living and working conditions. As Lowe (1990: 117) argues:

> the fiercest struggles in the redefinition of rural space were not with country people, who remained idealized but neglected, but between the conflicting recreational tastes and means of different urban groups, such as the *battue* of the plutocrats and *nouveaux riches*, the botanizing and rambling of the genteel middle class, and the hiking of the working class. The battles between these groups over rural

space were microcosms of their larger struggles for control over the urban social order.

Among the early groups to promote aspects of rural preservation were the Commons Preservation Society (now the Open Spaces Society) founded in 1865, the Society for the Protection of Ancient Buildings (1877), and the National Footpaths Preservation Society (1884). They all experienced difficulties when they began to challenge the established property rights of the private landowner, a conflict that the National Trust (founded in 1894) sought to resolve by combining an appeal to the notion of national heritage with a body empowered to buy and hold land for the nation. From the outset the NT fostered close links with large landowners. As Offer (1981: 342) mischievously points out, the NT thus became

a repository for the country-house heritage in Britain, which it acquired under a tenure of dual ownership which guaranteed the continued occupation of previous owners. The artifacts of a socially obsolete tenure were thus kept in being for their cultural value; and this was done by keeping a vestige of the old tenure. Thus were the rôles reversed, and tenure came to shelter behind culture.

Trust ownership by the NT was seen as a last resort but one accorded immense status by an Act of 1907 that gave it the power to declare its own land and buildings inalienable. The notion of inalienability had been central to the practice of entailment which had prescribed ownership of settled estates prior to the 1882 Settled Land Act. Thus preservationists began to assume the mantle of stewardship just at the time landowners were forced to discard it.

From the second decade of the 20th century onwards, rural landlords faced an almost inexorable decline in the extent of their estates. In 1909, David Lloyd George introduced his "People's Budget" which included proposals for higher death duties, site-value taxation and a tax on unearned increments in land values. It led directly to the constitutional crisis of 1911. The defeat of the Lords sealed the landed interest's loss of formal political power nationally. Although the taxation measures were in themselves modest, and the more radical proposals canvassed under the Liberals' Land Campaign of 1913 fell in the face of bitter opposition from vested interests and the need for unity on the outbreak of the First World War, the will of landowning families to assert their traditional interests was fatally weakened following the loss of innumerable heirs on the battlefields of north-eastern France. The rapid rise in wages during the war substantially raised the cost of maintaining a great estate, putting its proper maintenance beyond all but the most wealthy and self-indulgent. Moreover, the removal of agricultural subsidies in 1921 – they had been established in 1917 to aid wartime food production (see Whetham 1978) – reintroduced world market competition for British farmers, and dashed any lingering hope of a return to an agrarian-based rural prosperity.

The gathering class conflicts of the period had already stimulated the collective mobilization of competing agrarian interests, including the founding of the Central Land Association in 1907, the precursor to the Country Landowners' Association, and the National Farmers' Union in 1908. The establishment of the CLA reflected the new-found sense of insecurity felt by the rural landowner and landed capital, and the growing economic and political legitimacy of the tenant farmer. The NFU was established not only to demand greater security of tenure, and thus a more powerful position in relation to the rural landlord, but also to further the political influence of farming employers in the face of an increasingly organized and aggrieved agricultural working class. The war temporarily reduced such class conflict, and the food crisis of 1916 brought these bodies "into the arena of policy formulation and execution", providing them with a "raison d'être to their potential members" (Newby 1987: 164). It is an arena from which they have never withdrawn.

Relatively buoyant economic conditions prevailed in agriculture until the withdrawal of subsidies in 1921. Many indebted landowners took the opportunity to strengthen their financial positions by selling outlying estates and farms, and more than a quarter of all farmland changed hands in 1921–2. The major purchasers were tenants buying the freeholds to their properties, creating a new class of owners more attuned to the daily needs of making a living from farming. To many landowners of medium size and limited non-agrarian earnings, the landed estate had become an intolerable burden. In the light of what had gone before, "it is an irony of English rural history that the landowners who departed did so with as much relief as regret, happy to off-load an increasingly burdensome asset of doubtful value to those sufficiently optimistic to buy" (Newby 1987: 156).

The remaining landowners experienced the worst of the agricultural depression. They were obliged to find new ways of exploiting their estates as many of their tenants were by this time unable to muster the resources to buy their farms even had they wished to do so. A further adjustment in outlook was thus demanded, occasioning additional loss of social and physical exclusivity. The use of landed estates for country pursuits by a range of urban interests became an increasingly regular and commercial feature. But this also necessitated strict control over access that pitched landowners and their agents, especially in upland areas close to the industrial conurbations, against a tide of popular interest opposed to any notion of having to purchase access to the countryside.

In the 1920s, a rising standard of living, a shorter working week, the introduction of statutory holidays and improved public transport all contributed to an increased demand for consumption space by the urban working and middle classes. Demand could no longer be satisfied on those few isolated sites and properties held by charitable and public bodies, and a more comprehensive approach to the use and preservation of the countryside was called for. This brought into being new and powerful pressure groups that were divided between those that wanted to preserve rural areas from urban encroachment and those that wanted greater access to the countryside (Sheail 1981). Inevitably, tactics differed too. The Council for the Preservation of Rural England (CPRE), formed in 1926, lobbied for green belts and national parks, but like the NT fostered close links with landowning interests, seeing their revitalization as the main means of combatting urbanization. In contrast were those bodies whose main political cause was greater access (Youth Hostels' Association 1931, Ramblers' Association 1935) whose activities readily came into conflict with the interests of private owners. Their members, if not the bodies themselves, were prepared to confront private owners directly, as occurred most spectacularly in the mass trespasses and protest demonstrations on the southern Pennines in the 1930s. Together, these different pressures restricted landowners' options. They discovered, for example, that in exploiting economic gains from the upper middle classes in search of country pursuits, they became hemmed in, on the one hand by an urban middle class that often viewed the rural heritage as an asset to be preserved and not commercially exploited and, on the other, by a militant working class that demanded greater and free access to rural land for rambling and fishing.

At the same time, other more lucrative opportunities were opening up for some landowners through the rapid growth of suburbia, especially in the southern half of the UK, which made them key accomplices in and beneficiaries of urban development. Landowning interests had ensured that the limited planning controls over urban sprawl that were in place (e.g. in the Town and Country Planning Act 1932; Restriction of Ribbon Development Act 1935) were ineffective in practice and to their financial advantage. The level of compensation to which they were entitled for loss of development right under the planning legislation meant that:

> most schemes in fact did little more than accept and ratify existing trends of development, since any attempt at a more radical solution would have involved the planning authority in compensation it could not afford to pay. In most cases the zones were so widely drawn as to place hardly more restriction on the developer than if there had been no scheme at all. Indeed in the half of the country covered by draft planning schemes in 1937 there was sufficient land zoned for housing to accommodate 350 million people (Cullingworth 1974: 21-2).

The area under urban land use in England and Wales rose from 6.7 per cent to 8.0 per cent of the total in just nine years between 1931 and 1939, or by 200,000ha, a rate of growth not exceeded either before or since (Best 1981).

Landowners also went on the offensive in an attempt to recapture some of the high moral ground. In response to their increasing marginalization in national economic and political terms, they began to argue that they were the real custodians of the countryside and it would be to nobody's benefit were they taxed out of existence. Their case was undoubtedly aided by world economic and political events rather than any ability to reassert their former position in national life. The global economic recession of the 1930s and the decline of the Imperial order had exacerbated the existing difficulties facing the economy as a whole, as well as the farming industry. It led to some protectionism and increased acceptance of Keynesian views on what the state could achieve in the management of the economy. British farmers, and indirectly rural landlords, were to benefit from this change. Farmers began to receive modest subsidies on certain products and improved market regulation through the introduction of marketing boards (Ch. 3). These changes were designed to protect them to some degree from low and fluctuating world prices and from their weak bargaining positions with more powerful capital in the food industry (e.g. milk retailers). Before 1940, the impact of government support on incomes and wages was limited; it took the threat, and then the reality, of war for the government to put into effect a national strategy for *agricultural* development.

The food crisis and a changed perspective

The 1939–53 food crisis provided rural landowners with a renewed sense of national purpose. It offered the opportunity to combine material advantage with patriotic duty. In particular, the wartime emergency promoted state support for agricultural production, relegating the urge to reform landownership to the strategic need to produce. In the short term, this ossified property structures and slowed the rate of decline in landlordism. Indeed, there was very little change in farm size and land tenure during the crisis.

What the crisis revealed was a farming industry suffering from chronic underinvestment, often dreadful social conditions (see the returns from the *National farm survey* of England and Wales 1941–2, Whetham 1978, Orwin 1949), and extensive areas of derelict and semi-derelict farmland (Stamp 1950). It also highlighted the speed with which the farming sector could respond to financial incentives, albeit during a period of firm state direction of production targets and guaranteed markets. The sector was aided in this respect by a fully capitalist set of production relations and a large average size of farm which meant scale economies under an industrialized system of production could readily be realized. Expansion of output was also facilitated by a flexibility among landlords over tenants' rights, including a more liberal interpretation of the terms of their leases, and a political maturity among landowners who, with reluctance, accepted an increased degree of state control over the management of their properties in exchange for farm price support and the promise of higher rents when the emergency was over. While the onset of these new conditions undoubtedly compromised landowners, it also presented them with a new, productivist *raison d'être*. Although diminished in number, this allowed them to renew their claims to social leadership, both in the rural community and in national policy-making for the agricultural sector.

The central rôle that agriculture was to play in the postwar countryside was established under the broad terms of the 1947 Agriculture Act. The legislation effectively represented a contract between the state, farmers and landowners, and agribusiness, and in spite of periods of stress during the 1960s and 1970s, it held firm until the early 1980s. Notwithstanding the Labour Party's commitment to public enjoyment of the countryside, domestic economic circumstances and the international political economy of the late 1940s (see Ch. 3) dictated the utmost priority for food production. It is no coincidence that the 1949 National Parks and Access to the Countryside Act came towards the end of the first postwar government's legislative programme (see Cherry 1975, MacEwen & MacEwen 1982). By then, it provisions had to fit into a rural economy dominated by the needs of an agriculture backed by statute, encouraging a geographical separation between the farmed countryside and the conserved countryside. The conserved countryside was to be restricted to marginal farming areas, particularly in the uplands and wetland areas, thereby not interfering with the rapid improvement of lowland farmland. Even so, the creeping extension of intensive farming methods "up the hill" and the drainage and reclamation of marshes, wet meadows and bogs progressively encroached upon those areas designated for their conservation value under the 1949 Act.

However unrealistic this geographical separation might have been in the medium term, it was encouraged further by the nature of the 1949 Act itself, which created two relatively weak bodies. Responsibilities for access and amenity were separated from those for the conservation of wildlife, and were given to the National Parks Commission and the Nature Conservancy, respectively. Moreover, in its early years a scientific rather than a populist concern for nature dominated the thinking and practice of the Nature Conservancy, encouraging it to focus its attention on pristine examples of the nation's range of semi-natural habitats – frequently to be found at the margins of agriculture – rather than the regular habitat mosaics more typical of the countryside as a whole. It may have been persuaded to adopt this position by the underlying perspective of the Scott Committee's (1942) report on *Land utilisation in rural areas* which argued that amenity and conservation interests were safe in the hands of farmers.

Equally indicative of prevailing priorities was the compromise reached over public access. Under Part IV of the 1949 Act, certain public rights of way were given statutory force, thus providing the means to curb erosion of customary rights such as had occurred in the 1930s and 1940s, although the county councils, which were given the responsibility, showed little enthusiasm for the task. But the legislation failed to endorse earlier recommendations made to the government by the Scott, Dower and Hobhouse reports that the public should have a legal right of access to *all* open and uncultivated land. Full public access was not even to be sustained within the National Parks, on the reasoning of the Minister of Town and Country Planning that "where access was enjoyed by custom it was better . . . to let sleeping dogs lie and not to risk provoking landowners into withdrawing the 'privilege' of access. This was to be provided by law, as opposed to custom, only where an access agreement had been made with the landowner . . ." (MacEwen & MacEwen 1982: 19–20).

These circumstances left the private landowner not only with the right to exclude, but also the justification to do so on the basis of the food imperative and the need to protect from an intrusive public crops, stock and equipment into which more and more capital was being poured. The preservation of farmland was also achieved through a land-use planning system that required those who wished to transfer particular pieces of farmland to other uses to demonstrate that this was in the public interest and a planning ideology that paid considerable regard to the quantity and quality of farmland lost to other uses in any plans for urban expansion (see Ch. 5).

The primary concern of government, farmers and landowners was to harness private property rights to agricultural expansion. But by the 1970s, this exclusive position had begun to be challenged by a widening range of rural interests. For example, the 1968 Countryside Act was introduced in response to the vast growth in numbers of visitors to the countryside in the 1960s. They had to be catered for somehow and although the primary means advocated was simply to develop recreational enclaves, such as picnic sites and country parks, the 1968 Act did replace the National Parks Commission with the Countryside Commission. The new Commission, as its title suggests, had a much broader geographical and policy remit, and along with an increasingly vociferous environmental movement, became instrumental in raising public awareness of the destructive impact of intensive agriculture on valued landscapes and wildlife habitats (e.g. Westmacott & Worthington 1974, 1984). This evidence contributed to the introduction and stormy passage through Parliament of the 1981 Wildlife and Countryside Act, a process that brought fully into focus the root causes of the conflict between farming practice and environmental protection (Lowe et al. 1986). These concerned the socially fluctuating boundaries between public and private rights in rural land and the price "tags" placed on those rights through compensation payments to landholders for anticipated losses of profits arising from constraints imposed on land management practices for conservation purposes.

Compromising the landed interest: a productivist agriculture

As previously explained, during the period of the Imperial food order, agricultural production interests and the landlord–tenant system associated with it, had been sacrificed to larger economic and political objectives. The effect had been to weaken rural landlordism but not eliminate it. In 1940, over 60 per cent of farmland was still tenanted, the great majority in private owner-ship. The changes in farm policy from 1940 onwards challenged this dominant tenure more directly, by championing the effective use of productive capital. The rentier class was seen as a barrier to progress, with its control over a scarce resource exploited through the rent relation. Its influence has been progressively reduced. Since the mid-1950s there has been a continual shift from landlordism to owner-occupation, and from relatively simple to much more complex landholding arrangements, engaging not only landlords and tenants but also more members of farming families and even off-farm industrial capital.

Labour's election to office in 1945, committed to a set of radical policies, represented the greatest political challenge to landed capital. The government was opposed to the vestiges of the traditional tripartite rural social structure (landlord, tenant, hired agricultural worker), critical of the past lack of state control over land development, and committed (in its manifesto) to land nationalization. It did indeed substitute lifetime for annual tenancies, to give tenant farmers the necessary security to invest. However, the priority it gave to increasing food production in a period of shortage more acute than the war itself, provided a lifeline for private landlords. Choosing, in the event, to work with them and not to seek their abolition, the government put aside any determination to nationalize land (although it did nationalize development rights) or to abolish the tied-cottage system (Flynn 1986). It also compromised on its fiscal strategy, to provide rebates on income tax and surtax for capital invested in agricultural improvements: "Landowners with high taxable incomes could potentially convert former tax losses into real wealth by investing in land improvements" (Clemenson 1982: 113).

Most of the policies put in place between 1945 and 1950, which were already less radical than those outlined by Labour when in opposition, were further moderated in the ensuing decades. Nonetheless, conflicts over land policy between Labour and the Conservatives have been more evident in their treatment of the land-use planning system and the taxation of betterment (see Ch. 5) than in their policies towards agricultural land. Rhetoric aside, both have treated agricultural production as the key objective for rural areas and the decline of the landlord as a peripheral issue that could be allowed to take its course. Certainly, no government since 1950 has seriously considered introducing major restrictions on the ownership or transfer of freeholds. Instead, indirect means (fiscal and tenure legislation) have been relied upon to alter the pattern of occupancy and encourage the trend towards owner-occupation. Otherwise, the market has been treated as the most effective means of achieving an efficient distribution of freehold rights. Unlike several other European countries, no upper limit has been placed on the amount of land that a single individual or private body may hold or acquire, and purchasers of rural land do not have to demonstrate their proficiency in farming or estate management (for discussion see Northfield Committee 1979).

Thus the state has in practice opted to work with and through the existing ownership structure rather than engage in radical land reform. This compromise approach is a source of continuing tension between successive governments' general attempts to encourage agricultural accumulation and concern as to the justice of the outcome between, for example, landlords, owner-occupiers and tenants, or large and small producers. While state action has thus lacked consistency – preserving the autonomy of landowners on the one hand but then embarking upon sporadic political attacks on the landlord fraction on the other (Marsden 1986) – established landowning interests have managed to retain a position of influence in the countryside.

The state might claim that it has been successful in its indirect approach. After all, the land question (at least, in its strictly agricultural production sense) was suppressed as a major political issue until the 1980s, even if the details of tenure legislation have been contested at length by those representing the holders of existing rights. In addition, perhaps 75 per cent of farmland is now in *de facto* owner-occupation, and this relatively rapid change in occupancy has been achieved without unduly compromising agricultural production objectives. In particular, the growth in owner-occupation has been associated with a marked increase in the average size of full-time holdings since the mid-1950s in all tenure categories (Grigg 1987). Moreover, mixed tenure farms, combining freehold ownership and let land, are often among the more dynamic, expansionary and entrepreneurial of farming enterprises (Hill & Gasson 1985) (Fig. 4.1). The importance of historical circumstances continues to be revealed in the wide local differences in the tenure pattern today, even at the county scale (Fig. 4.2).

Comprehensive and accurate information on the changing pattern of rural landownership is almost non-existent, however, except at the most general of levels. The most recent reliable source on agricultural land is the report of the Northfield Com-

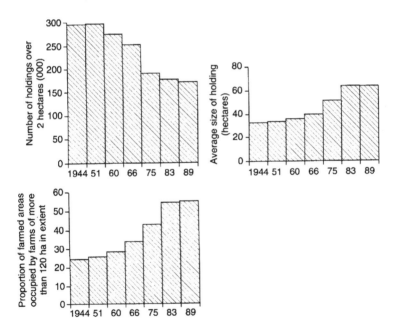

Figure 4.1 Change in farm size in England and Wales, 1944–89. (Source: *Agricultural statistics United Kingdom* 1989, HMSO, MAFF, London: a publication of the government statistical services.)

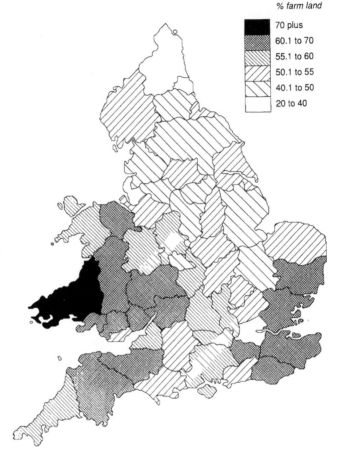

% farm land

70 plus
60.1 to 70
55.1 to 60
50.1 to 55
40.1 to 50
20 to 40

Figure 4.2 Population of farmland in owner-occupation by county, 1977. (Source: *Agricultural statistics united* (HMSO).)

mittee (1979). This states that in 1978, 1.2 per cent of farmland was owned by the financial institutions, 8.5 per cent by public and semi-public bodies and the traditional institutions (Church, Crown, universities, etc.), and the remaining 90.3 per cent (by subtraction) by private individuals, companies and trusts. No breakdown is offered of the last category, although the vast majority of owner-occupied land is included here, with let land dominating the first two categories. Except for some sample studies, information on the form of private landownership (i.e. land held singly, severally, by trusts, or corporately), or even the average size of ownerships, remains sparse. The most interesting longitudinal study of major private landowners is that of Clemenson (1982). She examined a random sample of 500 estates, all of which had extended to at least 3,000 acres in the 1873 *Return of owners of land*, and sought to establish their present size. Not surprisingly, over 90 per cent had declined in area but nearly half the families still owned more than 1,000 acres. The "great landowners" (owning more than 10,000 acres in 1873) had been more successful, or at least more tenacious, in retaining substantial estates. Clemenson's analysis indicates significant continuity of private ownership and the political influence at the local level that a large estate may still bring.

Even the substitution of owner-occupation for the landlord–tenant system has not always been at the expense of individual landowning families, for many of the larger landlords, encouraged both by taxation on rental income and production supports, have become large owner–occupiers instead. The 1987 Annual Rent Inquiry, for example, reveals that of the 4.5 per cent of all let farms that fell vacant that year, over half were not re-let and about one-third of these were taken in hand by the landlord (MAFF 1988). These forms of transfer are a potent reminder of a world where landlords are being encouraged to realize the full productive value of their assets and to treat their land as a source of fictitious capital.

During the 1980s, the financial institutions (represented by pension funds, property unit trusts and insurance companies) increased their share to approximately 2 per cent of all farmland, but with the onset of the farming recession and better investment opportunities elsewhere in the economy, their proportion has since declined sharply (Savills-IPD 1989). The modest growth in public ownership during the 20th century is attributable to a general expansion in state intervention in regulating the economy, and the need in specific instances to control the freehold to ensure a wide range of public policy objectives are met, rather than an attack on the private ownership of agricultural land. For example, following the widespread felling of timber during the First World War, the Forestry Commission was established in 1919 to create a strategic timber reserve; and water companies own land in upland areas to safeguard reservoir catchments.

Among public and semi-public bodies there has been some change since the 1970s. Amenity and environmental bodies have tended to increase their holdings while the disposal of some Forestry Commission land and parts of county councils' smallholdings estates has reduced theirs; other land has been lost to the public sector with the privatization of the water authorities. On balance, however, change has been slight and there has been no major challenge to the dominance of private ownership. This conclusion is supported by the data contained in the annual analyses of vendors and purchasers of farmland published by the Ministry of Agriculture, Fisheries and Food (MAFF). Only a small proportion of farmland is sold annually (about 1.5 per cent of the total in recent years) and in excess of 80 per cent of this is transacted between farmers (or agrarian landlords) with, on balance, a small net loss by them to other interests (e.g. developers, local authorities etc.).

The two most important sources of land mobility arise because of family changes – usually succession from one generation to the next – and the purchase of freehold land. The latter has become increasingly restricted, with less and less farmland sold each year. Its importance as a source of property rights has been enhanced, however, by the rapid decline in the number of new leases. Even until the 1950s, tenancies of different sizes still provided a valued means of advancement for ambitious farming families (Williams 1964, Nalson 1968). Farming families might have expected to remain in agriculture for generations, but not necessarily to occupy the same farm (Symes & Appleton 1986); today, the movement of tenants from one let farm to another is quite rare. Overall, the cumulative effect of high land prices, farm amalgamations and the very small number of new tenancies has been the creation of an industry of insiders. The purchase of land with vacant possession has become the means of entry for just a fortunate few, but is otherwise a major barrier to scaling the farming ladder.

According to information derived from the Agricultural Census, the proportion of farmland in owner-occupation was 38 per cent in 1950. By 1960, it had risen to 49 per cent as many landowners sought to rationalize their estates to cope with taxation and rising maintenance costs. By 1987, the proportion was 62 per cent, but that is almost certainly an underestimate of the amount of land in *de facto* owner-occupation. This is because occupiers hold a beneficial ownership interest in some of the land that is recorded as rented, as is often the case with family trusts and companies. If it is assumed that almost all the land held by the financial institutions and public and semi-public bodies is let, then the amount of land let at "arms-length" by private owners may now be less than one-fifth of the total area compared to four-fifths in 1910.

The growth in owner-occupation since 1950 has been primarily instigated by legislation designed to strengthen the rights of tenants and by a fiscal system that has penalized landlords more than owner occupiers. In an attempt to protect the tenant's interest and to encourage investment in production, the first postwar Labour government passed the 1948 Agricultural Holdings Act. Under its terms, annual tenancies were replaced by lifetime tenancies, provided the occupier maintained acceptable standards of estate husbandry and paid the rent. Rent levels were subject to independent arbitration if the landlord and tenant could not agree and this is still the case. By and large, this arrangement has worked reasonably well even if it encouraged some landlords to sell up in the 1950s.

A much greater affront to landed interests was caused by the 1976 Agriculture (Miscellaneous Provisions) Act which permitted the extension of secure occupation by the farming family to three generations provided fairly modest eligibility (close kin) and suitability (experience, age, financial resources) tests could be met by successors. It effectively gave primacy to the needs of existing farm families rather than any need to revitalize the agricultural sector through helping outsiders to enter the industry. Opponents of the legislation claimed that it would hasten the demise of the dwindling let sector, and thus turn farming into a closed shop. The available evidence tends to support this claim and in 1985 the Conservative government sought to return to the single life tenancy created by the 1948 Act. As yet, there is no evidence that this change in policy has had the desired effect of reversing the growth in owner-occupation. Most landlords now so mistrust the legislators that they are unprepared to re-let farms, and seek to protect their own long-term interests either by farming the land in-hand or in partnership with farming companies.

Mistrust over future tenure legislation is compounded by the other area of disadvantage for landlords: taxation. In terms of revenue, for example, rents are treated as unearned income and are therefore subject to much higher marginal rates of tax than farmers' incomes. A similar distinction arises where capital taxation is concerned, although in an effort to encourage them to invest in food production, landlords were able to claim, after 1948, a 45 per cent abatement on the evaluation of land for Estate Duty (inheritance tax). Furthermore, provided the estate was passed on at least seven years before the owner's death, Estate Duty could be avoided altogether. Subsequent changes to capital taxation rules have made the position extremely complex but on the whole they have maintained a favourable status for the owner–occupier (see Stanley 1984, Northfield Committee 1979), further discouraging the letting of land on secure leases.

What is less frequently examined is whether the fixing of large sums of capital in freehold ownership has drained the industry (or even the national economy) of capital that would have been better invested in farm business expansion, and whether this process has not simply helped to create a new group of self-selecting insiders dependent more on their familial position than their ability to farm. This conclusion may do some injustice to the market's ability to weed out the inefficient producer. Nonetheless, family succession and continuity among the financially adept have been reinforced during the productivist period, creating both a social as well as a capital fixity in resources. The creation of a secure market for agricultural commodities encouraged a strong rise in farmland prices in real terms during the 1960s and 1970s. And these were increased further by the inflationary pressures of the 1970s and the growing demands of non-agricultural uses for rural land. These developments took them back to levels higher than those of the 1870s and more than five times the level during the interwar recession (Fig. 4.3). High land

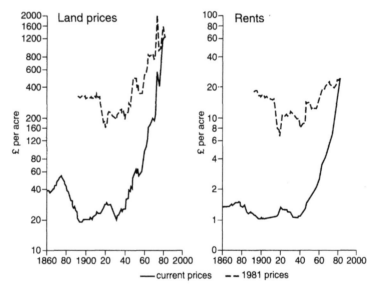

Figure 4.3 Changes in agricultural land prices and farm rents, 1860–81. (Source: based upon Jones Lang Wootton 1983.)

values provided collateral, but borrowing against it contributed to a much more heavily indebted industry during the 1980s and 1990s as incomes fell and interest rates rose. The level of income gearing – that is the ratio between the sums spent on the repayment of debt and net farming income – has often been in excess of 35 per cent in recent years.

Finally, within this broad picture there has been a growing diversity of landholding arrangements, designed both to reduce taxation liability and to maximize the opportunity to accumulate. In the case of let land, there has been a widespread and consistent increase in the amount of farmland rented on insecure leases. More than a quarter of all let farmland may now lie outside the standard terms laid down by the Agricultural Holdings Acts (Winter et al. 1990). Within the category of owner-occupation, there have been major changes to the arrangements affecting the detailed distribution of property rights and control over their useand exchange. Except in the more conservative farming areas, the traditional sole-operator form of owner-occupation has become a residual category. The dominant forms involve several members of the family and, in the urban fringe, representatives of industrial and banking capital as well (Whatmore et al. 1990, Munton et al. 1988). More generally, banking and finance capital have entered into a widening set of arrangements with farm businesses (usually indirectly via various forms of credit), becoming not only the primary lender for short-term needs (e.g. overdrafts) but also for medium- and long-term investment.

The reasons for these changes have already been identified. On the supply side, they reflect an increasingly desperate search among farming families to establish new business opportunities on their farms and to raise capital by treating their land (or its buildings) as fictitious capital. On the demand side, the excess speculation in agricultural land of the 1970s, attributable to high rates of inflation and limited investment opportunities in the rest of the economy, has been dampened. The financial institutions, which threatened to become the key new investors in the 1970s, have therefore been replaced by a wider range of interests with a more direct concern for the countryside, either as a place to live or to develop a business. The pattern of these demands is very uneven spatially, and has varied temporally during the 1980s as successive Conservative governments have struggled to achieve a politically acceptable compromise between their free-market, development-oriented wing and those among their traditional rural supporters committed to rural protection (see Ch. 5).

the value of the agricultural asset base continues to be reflected in the uneven price of farmland (much higher on small units including houses, especially in south-east England) and in the contribution of "residential buildings" to the balance sheet of British agriculture. At the end of 1988, such buildings were worth £9.9 billion, or 15 per cent of all the industry's assets, including land (Johnson 1990).

Compromising the landed interest and protection of the rural environment: from custodialism to compensation

Given the priority attributed to food and, to a lesser extent, fibre production, in the postwar period, it is no surprise that up until the late 1970s most of those with extensive rural property rights (owner–occupiers as well as landlords and tenants) associated their economic wellbeing with these enterprises, and often did so with singular disregard for other interests. They had yet to be fully alerted to the emerging agricultural crisis, or the growing market for traditional country pursuits, such as shooting, fishing and riding, or the expanding demand for golf courses, orienteering facilities and the acquisition of "out-of-bounds" skills. These frequently remained beyond the market, either as exclusive luxuries for the landed wealthy or identified, in part, with public sector provision. They had yet to be seen as new commoditized rural goods.

More significantly, these same property interests frequently rejected the claims of environmentalists that they were harming the beauty, traditional character and wildlife content of the countryside, in spite of mounting evidence to the contrary (e.g. Westmacott & Worthington 1974, Nature Conservancy Council 1977, Perring & Mellanby 1977). They were roused to anger by those who not only asserted public environmental rights on private land but also dared to challenge the very basis of the private ownership of land (e.g. Shoard 1980, 1987, Norton-Taylor 1982). They were particularly affronted by the argument that they were damaging the countryside through the use of public moneys, whether in the form of price supports, capital grants or tax reliefs (Bowers & Cheshire 1983, Jenkins 1990).

In their defence, rural property interests chose to raise the matter to one of principle, although it was hardly a new principle in the long-standing debate over the relations between private property and the state. They viewed the environmentalists' challenge as fundamental in two, linked regards. First, they treated it as a threat to their freedom to manage, whether dressed up as management agreements or as state-imposed regulation of management practices; Secondly, they objected to any such imposition, were it not to attract adequate compensation to offset any loss of income. Such losses could arise either from the additional costs incurred in managing land in a prescribed way, or from income forgone because the land could no longer be used in its most profitable manner. Refracted through these twin preoccupations, political demands for the reform of agricultural policy and the regulation of farming practices to protect the countryside have been translated into landowners' demands for the establishment of appropriate market and support mechanisms to recompense them for the provision of environmental "goods".

Initially, the representatives of farmers and landowners fell back on the custodial arguments that had served their freedom of action and accumulation interests so well during the productivist period. They emphasized farming's contribution to feeding the nation and traditions of stewardship in managing the countryside (National Farmers' Union & Country Landowners' Association 1977). It was argued that meeting external pressures for environmental protection could jeopardize both these objectives. But as public and political opinion moved more emphatically in favour of stronger environmental safeguards, a more active defence of farming and landowning interests was called for to ensure that any unavoidable restrictions did not disadvantage them. Broadly, their campaign had four objectives:

o A minimization of the spatial extent of any new controls.
o A guarantee that policies would be permissive (i.e. controls would not be mandatory, acceptance of them would be voluntary and their terms would be negotiable).
o A guarantee that compensation for financial loss should be paid at full market value on the basis of annual payments.
o A guarantee that controls or agreements should be negotiated with MAFF and its agencies, or those bodies with which farmers and landowners had had a long-standing corporatist relationship (Cox et al. 1986a,b). If the industry were not to be allowed to regulate itself, the dead hand of an ill-informed bureaucracy should be avoided at all costs. Controls would need to be sensitive to local circumstances, it was argued, and their implementation would only be effective if agreed by those familiar with the economic and estate-management

problems of farmers and landowners. Only then might the latter have any confidence in the regulatory system.

The challenge to rural landowners arose initially at a local level, linked to a series of specific disputes from which farmers and landowners did not emerge with credit, such as the reclamation of moorland on Exmoor and the drainage of parts of the Halvergate Marshes on the Broads (Lowe et al. 1986). These local disputes provided the opportunity for national pressure groups to transfer the struggle over the future of public policy to Whitehall and Parliament, at which point the focus of attack was the nature of public policy rather than the individual accumulation strategies of farmers and landowners. The farming lobby sought to deflect the environmentalists' case by advocating improved advice for farmers and a voluntary system of management agreements, and by advertising their active participation in the Farming and Wildlife Advisory Groups – an organization committed to promoting conservation among farmers (Cox et al. 1990).

Simply getting MAFF to accept the need to build environmental considerations into agricultural policy and then into agricultural funding was a major task. In spite of its crucial rôle, MAFF had not appeared to take much notice of the responsibility placed on it by Clause II of the 1968 Countryside Act. This required all government departments "in the exercise of their functions relating to land" to "have regard to the desirability of conserving the natural beauty and amenity of the countryside". The word "desirability" left too much to interpretation. For example, in its 1974 White Paper, *Food from our own resources*, MAFF made no mention of environmental considerations in its drive to expand output. Its initial acknowledgement, and that of the NFU, of the conflict between agricultural objectives and amenity was to restrict it to "maverick farmers" and to areas of special conservation value. It was not a conflict intrinsic to industrialized farming practices and agricultural policy.

The debate that above all others concentrated public attention on these issues occurred during the passage of the 1981 Wildlife and Countryside Act (see Cox and Lowe 1983, Cox et al. 1985). It provided the opportunity for environmental interests to press for public environmental rights over private property. They faced an uphill struggle. Ministers and their officials sought to limit the scope of the legislation to a series of minor reforms pertaining to the conservation of semi-natural habitats and open moorland in National Parks. Moreover, the terms of the Act, *sensu stricto*, and their scope were little altered during parliamentary debate, and in the area of property rights the Act is generally regarded as a triumph for landowning interests. They were able to mobilize greater resources than the conservation bodies at all stages in the parliamentary process, constantly reiterating a clearer and more consistent position than their opponents. With only one modest breach, they were able to see the triumph of their principles prevail: of voluntary cooperation, encouraged where necessary by management agreements based upon full financial compensation; and a focus of concern limited to relatively small parts of the countryside of high amenity and conservation value.

At first sight, conservation interests appear to have been the losers, and continuing evidence of decline in amenity and wildlife resources seems to bear this out. Maintenance of the voluntary principle under the Act means that farmers cannot be prohibited from intensifying the exploitation of their land, even if grant aid in support of a development scheme can be withheld. The ability of property owners to maintain the principle of compensation was, perhaps, their most astounding achievement, especially as compensation is paid to offset hypothetical income forgone. As the then director of the CPRE exclaimed in a letter to *The Times*: "It gives legal expression to the surprising notion that a farmer has a right to grant-aid from the tax-payer: if he is denied it in the wider public interest, he *must* be compensated for the resulting, entirely hypothetical 'losses'" (original italics; quoted in Lowe et al. 1986: 147). This principle also means that the subsidy paid from one area of public spending (agricultural policy) inflates the level of compensation to be paid from another (environmental policy). It was only later, with the introduction *inter alia*, of Environmentally Sensitive Areas in 1985 and Nitrate Sensitive Areas in 1990, that compensation payments for environmental objectives could be drawn directly from agricultural funds. The ability to "transfer" funds from food production to environmental protection – an unreal distinction between functions given the fact that agriculture is integral to both – has now received formal blessing from the European Community. Indeed, contemporary reform proposals for the Common Agricultural Policy include measures to encourage farming methods that enhance, rather than diminish, the quality of the rural environment.

Conservation interests made some progress, therefore. Before the 1991 Act the onus had been on them to demonstrate the damaging effects of changing farming and forestry practices, but subsequently the agricultural community and the government had to ensure the Act's effectiveness in halting the destruction of wildlife habitats and landscapes, especially if further legislation is to be avoided (Cox et al. 1985). Further, the debate on the Act allowed the expression of more fundamental objections to agricultural policy, including the case for the 'cost' of conservation being borne by the much larger agricultural budget. The prospect, moreover, of voluntary or involuntary disinvestment from farming, through a decline in the economic position of farmers, raises the possibility of achieving certain environmental objectives associated with deintensification, rather than through the imposition of controls. At least now these and other options can be actively debated because the 1980s were characterized by a small but consistent withdrawal of fixed capital from the farming industry, once depreciation of the existing stock is allowed for, reversing the trend of the previous 40 years (Fig. 4.4).

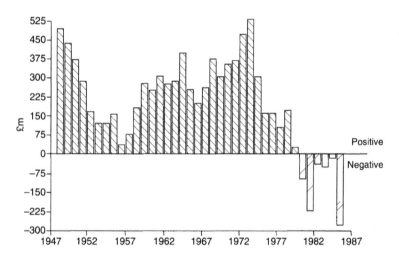

Figure 4.4 Year-on-year changes in the fixed-capital stock of the UK farming industry, 1948–86.

Conclusion

Our analysis indicates that the postwar transition to a productivist food regime, like the more recent ascendancy of environmental and other consumption interests in rural land, challenged established landed interests but failed to eradicate their influence. Many landlords have been induced to follow the lead of many former tenants and join the burgeoning ranks of owner–occupiers. With the professional aid of land agents and farm managers, they have successfully embraced agricultural productivism. Those that have continued to let their land have had to keep a watchful eye on changes to fiscal and tenure policies, which, on the whole and subject to individual circumstances, have encouraged them to diversify the types of legal control they hold over land. They have taken full advantage of the divisibility of rights available under freehold ownership, leading most recently to a rise in short-let land. Fixity of capital and inflexible distributions of property rights are not favoured by prevailing economic and political circumstances.

The most compelling message to emerge is the ability of landed interests to alter the representations of their position to take account of changing circumstances, and to do so effectively in a context of declining economic and political power. The extent of their landholdings and the form that they now take may be quite different from those of 1890, but the ability to influence the pattern of rural development remains. At the end of the 19th century, landownership was represented as a custodianship of rural society and the nation's heritage; a similar argument, with reference to the environment, has been made with mixed success in recent decades in an attempt to refute the charge that modern farming practices are damaging to the countryside. During the productivist period, existing owners sought to argue that only those with a long and detailed knowledge of the land could ensure its efficient but caring management, and this has formed part of their case over many years that the level of inheritance tax should be reduced and that new agrarian landlords such as the financial institutions should be kept out (Munton 1984). Family continuity and insider knowledge were paraded as essential to a stable and sustainable rural economy, even if owner-occupation rather than the landlord-tenant system was the preferred tenurial arrangement. The shift to owner-occupation itself is also symptomatic of a commoditization process that has seen the substitution of landed capital by industrial capital, the payment for environmental goods that were formerly viewed as customary, and even the increased involvement of finance capital in an industry that now carries a heavy debt burden in relation to its cash flow if not to its net worth (see Fig. 4.4).

The price of the industry's primary capital asset, its land, does not reflect the economic position of farming in the national economy or in the food system. In both cases, farmers contribute an ever decreasing share of the value added. It records instead the widening demands placed on rural space by competing consumption interests. Thus as the productivist regime has declined, landowners and their representatives have turned to the maintenance of their freehold rights: especially the potential threat to their freedom of action represented by extensions to environmental and planning legislation. Having found it difficult, but not without some financial reward, to sustain their claim to the moral high ground of environmental custodianship, it has been the creation of new and diversified markets beyond agriculture that has beckoned. This prospect represents the most obvious source of future accumulation but, as we shall see, requires owners to take risks in unfamiliar and less protected markets, and provides a very uneven pattern of opportunities. We now turn to the changing context in which rural land development has been regulated during the 1980s, and within which landed interests, along with other actors, have had to represent and negotiate their positions. At the minimum, landed interests retain a "power of constraint" in the processes of change, but many adopt a much more proactive stance.

CHAPTER 5

Planning and the rural land development process
The reconstitution of the public interest

Introduction

The planning system is one of the critical axes of the land development process. At a procedural level it prescribes the rules and regulations through which most, but not all, land development must pass. At a sociological and political level it provides a means to mediate local conflict and a focus for it. It is therefore a key arena for the representation of interests, especially since the encouragement of public participation in the planning system in the 1960s. The election of the Conservative government in 1979 did not undermine the mediating rôle of local government, but shifted its focus. Ideas of positive planning have subsequently largely been undermined and increasingly local authorities have found themselves acting in a responsive manner and regulating private sector led development.

In other ways, however, the Conservative government of the 1980s proved to be a watershed for both agricultural and planning policy. Whereas previous postwar administrations had successively amended their objectives in these two fields, the Thatcher administrations challenged the very basis of policy. From a New Right perspective, the statutory planning system was regarded as a brake on market forces, and agricultural supports were regarded as an unacceptable form of corporatist sectoral management. The government's specific stance towards rural land development issues was also shaped by shifts during the 1980s in its core political support in the countryside from farmers and landowners to the "service class". As private interests in rural land became more diversified, so the public interest – expressed through the planning system – was seen to be in need of considerable reorientation.

Despite the radical intent of the Conservative government elected in 1979, there is a tendency for commentators to accept rhetoric for action uncritically, and to fail to set the reforms of the

1980s in the broader sweep of postwar change. Within the planning field, for example, the government's attempts to foster market forces are but part of a longer-term retreat, with limited and temporary exceptions during subsequent Labour administrations, from the initiatives of the 1945 Attlee government. Those early postwar years were marked by a widespread recognition of the urgent need to regenerate the nation's urban infrastructure and basic industries in the wake of the destruction and dislocation wrought by the hostilities. Drawing on strong feelings of national unity and purpose instilled by the war effort, considerable faith was placed in the ability of the state to plan and orchestrate regeneration, in deliberate contrast to the laissez-faire approach to urban growth and industrial development of the interwar years. Indicative of contemporary thinking were three wartime reports (Scott, Barlow and Uthwatt) on how urban and rural society might be organized when peace came.

A key rôle was envisaged for a comprehensive system of land-use planning that would not only guide but also lead regeneration. An obstacle to such an approach in the pre-war years had been the necessity to compensate landowners for restrictions imposed upon their rights to develop their land, but under the terms of the 1947 Town and Country Planning Act, development rights were nationalized. Landowners who could claim pecuniary disadvantage as a result – that is, could reasonably have expected to profit from development at that time – could make a once-and-for-all claim on a compensation fund of £300 million established by central government. In the future, the right to decide which land could be developed, and for what use, was to be lodged with local planning authorities, and compensation could not be claimed for refusal of planning permission. Moreover, local authorities, government departments and statutory undertakers could compulsorily acquire sites for development at existing use values, whereas the private initiation of development by the landowner would attract a betterment tax at a marginal rate of 100 per cent, known as the Development Charge, on any development gains.

By these means, landowners' rights were significantly curtailed, but not to the extent of their worst fears. Other private property rights were not nationalized and, of crucial significance subsequently, landowners were left with a rôle in the development process, initially through default and poor logic, then by design. Outside of the new towns, neither the new Ministry of Town and Country Planning nor the new local planning authorities were given the financial resources they needed to promote the comprehensive social and economic changes they wished to pursue through land development. Most of the financial power remained with other ministries with sectoral interests, such as transport, energy and industrial development. The Act only provided a *framework* within which positive, state-led planning could occur; functionally, the planning system remained in a regulatory mould, increasingly able only to determine those changes in land use that fell within the definition of development. The system became increasingly dependent on private interests to achieve its publicly stated goals (McKay & Cox 1979, Ambrose 1986, Reade 1987). For example, when the Conservatives returned to office in 1951, they took steps to undermine the public sector's lead rôle in development. First, they removed the Development Charge, leaving all the development gains in the hands of landowners, thereby encouraging land to be brought forward through the private sector. Secondly, in 1959, market values were substituted for existing use values as the basis for payments for land acquired compulsorily by public bodies.

With the repeal of taxation on betterment, the planning system ensured large and certain capital gains for those owners whose land was identified for development under the local plan. In other respects too, the planning system has accorded the landowner, whether the original owner or the developer, considerable negotiating powers. For example, as outline planning permissions are normally valid for three years, timing of the sale and development of the land lies largely in the hands of private interests. Moreover, those whose land is not identified for development are not prevented from submitting planning applications. Only recently has a charge been introduced in the procedure for making applications, and an additional amount levied where an application not in accordance with the local plan is refused. Similarly, the landowner's negotiating position is reinforced by the right of appeal to central government in the event of a refusal by a local planning authority, a right that can place pressure on local authorities to reach an accommodation with the applicant to avoid any loss of local control over the outcome (Goodchild & Munton 1985, Cloke & Little 1990). The ability of the private sector to negotiate its interest is no better illustrated by the amount of land that is developed outside those areas identified in local plans and even in those where there is a nationally acknowledged presumption against development, for example in Areas of Outstanding Natural Beauty (Anderson 1981), National

Parks (Brotherton 1982) and Green Belts (Munton 1983).

Even the radical intent behind the Labour government's Community Land Act of 1975, which proposed to take into public ownership all development land identified by local planning authorities as necessary in the public interest, provoked a remarkably effective fight-back by those representing private property interests. Through a succession of legal disputes, the legislation was largely emasculated by the courts. In retrospect, the 1975 Act can be seen as the last major effort to return the postwar planning system to its original conception, under which the allocation of land uses was divorced from the calculations of private landowners. Since then, the emphasis has been on leaving allocation to the market, and hence the initiation of development to the private owner, subject to the regulation of the statutory planning system.

The 1947 Act did not simply that it help set the parameters of the debate on the rights of landowners, it also established most of the instruments of the planning system. These instruments remain remarkably intact (Fig. 5.1), which perhaps serves to reveal some of the broader interests of landed property in the certainty and security of a regulated land development process. Healey and her colleagues, in their review of the pressures the planning system faced in the 1980s, emphasise the degree of procedural continuity. "Despite some current political rhetoric," they argue "it is not the existence of a programme for managing land use change which is under fundamental challenge. Rather, it is the *policies* pursued through the system and *practices* by which it operates which are being questioned" (Healey et al. 1989: 10 [emphasis in original]).

It remains open to question, therefore, what the real impact has been of minister's neo-liberal radicalism by the time that it has been filtered through the devolved structures of British administrative government (Rhodes 1986, 1988). It is a question that can be settled ultimately only by empirical enquiry and analysis. Although there have been a large number of studies of urban planning and land development in the 1980s (e.g. Rydin 1986, Goodchild & Munton 1985, Healey et al. 1988, Ball 1985, Brindley et al. 1989), rather less research has been undertaken on the rural land development process (but see Short et al. 1986, Barlow & Savage 1988, Cloke & Little 1990) despite the fact that it has encountered quite distinct economic and political pressures.

Type of land policy measure	Specific instruments	1945–51 LABOUR	1951–59 CONSERVATIVE	1960–64 CONSERVATIVE	1964–70 LABOUR	1970–74 CONSERVATIVE	1974–79 LABOUR	1979– CONSERVATIVE
REGULATORY MEASURES	Control of development	1947 →						
	General development order	1947 →						} major revisions
	Use classes order	1947 →						
	Rights of consultation	1947 • public agencies and government departments				1971• publicity for some types of development →		
	Rights of objection	1947 • direct property →						
	Public inquiries	1947 •						
	Regional policy controls	1945 • Industrial Development Certificates →			1965• Office Development Permits →			
	Special area provisions	1949 • national parks and areas of natural beauty →; 1947 • comprehensive development areas →			1967• conservation areas →; 1968• general improving areas (housing) →		(industry)	1980• enterprise zones; 1986• simplified planning zones
DEVELOPMENTAL MEASURES	Land acquisitions for planning purposes	1947 •	1963 • by agreement for any purpose →			1972 • for some planning purposes →	1975 • Community Land Act (for public purposes)	
	Compulsory purchase for planning purposes	1947 • for planning purposes →						
	Powers for planning purposes							
	Special development agencies	1946 • New Town Corporations →; 1947 • Central Land Board →			1967 Land Commission •		Most NTCs being 'wound up'; 1980 Urban Development Corporation 1980	
	Compulsory disposal of public land							
FINANCIAL MEASURES	Taxation of development gain	1947 •— Development charge — (1952)–1953			1967• Betterment Levy • 1974 Development gain tax	1976 1976• Development land tax →	• 1985	
	Public purchase of land and property at a privileged price	1947 •	• (1959)					
	Planning agreements				1971 Section 52 →			
	Subsidies for land reclamation; for development				1966• Derelict Land Grant →	(reclamation for development emphasised)	1980• Urban Development Grants; 1980• Enterprise Zones (financial provs); 1986• Urban Regeneration Grants	
INFORMATION AND GUIDANCE	Development plans	1947 • 'One-tier' plans →			1968• structure and local plans →	— until revoked —	(major revisions discussed); 1980• registers of public land holdings; 1985• unitary development plans	
	Public inquiries	1947 •			(local plans only) →	1972• examination-in-public for structure plans →		
	Consultation procedures	1947 • for government departments and public agencies →			1968• with the public at large in respect of plans →			

Figure 5.1 Instruments of the English planning system. (Source: Healey 1988.)

The Conservative government's neoliberal approach, for example, has been mediated by three specific factors when applied to rural areas. First, the countryside remains the Conservative heartland and therefore rural policy has been more receptive to traditional shire Tory interests; market liberalism has remained respectful of rural propertied interests. Secondly, agricultural production has demanded pragmatic responses within a supranational policy framework that is fundamentally welfarist, Keynesian and interventionist. Thirdly, the government has become increasingly sensitive to a rising tide of popular concern for the environment; often most forcefully expressed within a rural context (Lowe & Flynn 1989).

It is against this background that we examine the evolution of rural planning in the UK since the 1970s. We outline first some of the key elements of Conservative planning policy more generally before turning to their rural dimensions. The consequences of policy changes are examined by concentrating upon rural land development, with an indication of some significant differences between the rôle of public planning in the urban and rural domains. As we have argued in Chapter 3, the rise of a post-productivist phase of agricultural regulation also raises important questions about the rôle of the land-use planning system in rural areas.

Conservative planning policies

From the outset, the Conservative administration elected in 1979 indicated its intention of reducing the scope of the planning system by, for example, repealing the Community Land Act, thus undermining the positive rôle of local authorities in the development process. The scrapping of Regional Economic Planning Councils in 1979 curtailed strategic planning, and the subsequent abolition of the metropolitan counties and the Greater London Council severely reduced the basis for conurbation-wide planning. Moreover, local planning authorities have regularly been urged to restrict their attention to matters strictly relevant to physical planning and not to pursue wider social and economic objectives through the planning process.

Alongside these efforts to narrow the scope of planning have been steps to remove some of the constraints faced by developers. The Use Classes Order (UCO) and the General Development Order (GDO) are two of the major instruments that operationalize the planning acts by specifying those types of land-use change and development that are either within or outside the control of local planning authorities. The GDO is a permissive measure that allows a wide range of development to take place without planning permission. Among the principal areas of "permitted development" are those relating to agriculture and forestry. The UCO, on the other hand, categorizes the uses of land and buildings, and offers freedom from planning control for certain changes of use *within* a particular land-use category. The broader the classes under the UCO, and the wider the scope of the GDO, the greater is the freedom offered to landowners and users to utilize land and buildings in alternative ways. Both instruments have therefore offered scope for the Conservative government to pursue its liberalizing strategy.

Initially, it directed its attention towards reforming the GDO, amending it in 1981 and making further changes in 1985. In general, these regulatory reforms, that is those that have not followed from new, primary legislation, have been quite modest. There has been some updating of the GDO, some specific adjustments to accommodate other reforms (for example, extending to privatized utilities and Urban Development Corporations permitted development rights traditionally enjoyed by statutory undertakings and local authorities respectively), minor relaxations to aid industrial or commercial expansion, and some freeing of the scope for home extensions.

The government's approach to the reform of the UCO in 1987 proved more radical and was described at the time as "the biggest change to the planning system since the public participation and conservation measures of the early 1970s" (Home 1987: v). This followed the 1985 White Paper *Lifting the burden* (Secretary of State for the Environment 1985), which had attacked the levels of bureaucracy and delay within the planning system. The most significant amendment was a new business class created by combining the two classes on offices and light industry. Another change created a combined class for assembly and leisure use, whose particular implications for the countryside are considered below. A year later, in 1988, a new GDO was published under which most of the substantive changes related to rural land (Grant 1990).

As well as these efforts to liberalize planning control, there have been parallel attempts to promote the rôle of private capital in the development process. The Conservatives' first move in this direction was the introduction in 1980 of Enterprise Zones

(EZ) where developers would benefit from streamlined planning procedures, as well as tax exemptions. Later, in 1986, Simplified Planning Zones (SPZ) were introduced that allowed developers similar freedoms from planning control but without any financial inducements. Reliance on the private sector as the agent of urban regeneration is a central feature of the government's inner-cities policy. Private developers have been encouraged to take advantage of the enforced sale of land owned by local authorities and of relaxed planning controls in order to replace the traditional local authority rôle in land assembly and site redevelopment. To catalyze this shift, Urban Development Corporations (UDC) have been established in a number of inner-city areas, and, more recently, a national Urban Redevelopment Agency. All are agencies of central government vested with powers over the use of public land, compulsory purchase powers and responsibility for development control.

More generally, planning authorities have been repeatedly exhorted by ministers to take a much more sympathetic and accommodating stance towards development interests. The White Paper *Lifting the burden* observed that the planning system "imposes costs on the economy and constraints on enterprise that are not always justified by any real public benefit in the individual case" (para. 3.1). There was therefore a need to "simplify the system and improve its efficiency and to accept a presumption in favour of development" (para. 3.4). Such prescriptions were reiterated in several government circulars, and ministers signalled their intention to reinforce this advice through the decisions reached via the appeal system. The proportion of appeals allowed rose sharply during the 1980s and this in turn encouraged more developers to appeal. Between 1983 and 1988/89 the annual figure more than doubled to 25,000. Inevitably, this placed a considerable burden on central decision-making. Significantly, the number has not risen further (DOE 1992a, para. 14), but this may also reflect the impact of economic recession on the development industry.

The government also sought to streamline the local planning system and to speed up its decision-making, culminating in the consultation paper *Efficient planning* (DOE 1989a) issued in July 1989. The bulk of its recommendations, however, were aimed at reducing the burden of planning appeals on central government by, for example, introducing fees for those who appeal against decisions, and by giving the Secretary of State the power to deal with an appeal by written representations rather than going to a public inquiry. Other proposals sought to prevent repeat applications by developers that had already been rejected on appeal, and to reduce the number of planning applications that actually come to appeal by encouraging local authorities to become more accommodating to developers. Already, the Secretary of State possessed powers, which have increasingly been exercised, to award costs against local authorities which, on appeal against a refusal of planning permission, are deemed to have acted "unreasonably".

The reform and reorientation of the planning system have been most vigorously pursued by the government in the major conurbations and industrial cities (Brindley et al. 1989). Not only have these urban centres lost the strategic planning functions held by the former metropolitan authorities, but they have also been the recipients of EZs, SPZs, UDCs, and so forth. Such deregulatory initiatives embodied not only the neo-liberal impulse of Thatcherism, but also its more authoritarian side. This was stimulated by the imperative both to promote urban renewal by the private sector in the wake of inner-city rioting in 1981, and to defeat municipal socialism. Indeed, one effect of the urban planning reforms imposed by central government has been to curtail local authority control and opportunities for public participation generally. Other changes to the planning system, such as relaxations to the GDO and the UCO, similarly involve both liberalization and a diminution of local democratic control.

This centralizing tendency was also explicit in the abolition of the metropolitan county councils in 1984. Their co-ordinating and strategic planning functions were superseded by the Secretary of State's strategic guidance and by his rôle as the sole arbiter between competing demands from neighbouring authorities. Blowers (1987: 284) has commented:

> The shift from strategic to local planning is favourable to private developers who can bring direct pressure to bear on district councils responsible for development control. At the same time the simplification and greater central control exerted through the Secretary of State favours the big developers who enjoy considerable influence at the national level.

Similar forces can be seen at work in the revival of a form of regional planning. As the government put it: "there are issues which, though not of national scope, apply across regions or parts of regions and need to be considered on a scale wider than a single county or district" (Secretary of State for the Environment et al. 1990: 86). Although local authorities were to initiate the process by producing draft guidance, it was left to central government formally to issue regional guidance after engaging in its own consultation procedures.

Rural authorities, of course, have not been unaffected by the general liberalization of the planning system under the Conservatives, although in areas of high environmental value some controls have actually been tightened or additional safeguards introduced (see below). In general, the government's enterprise culture rhetoric and its emphasis on small businesses and flexible working practices has meant that it sees rural areas as providing significant development opportunities (Boucher et al. 1991). Efforts to streamline and relax the planning system were pursued as much in rural areas as in urban areas, but this liberalizing thrust was not reinforced by the interventionist and centralizing measures deployed in urban areas. It did not therefore undermine local democratic control to the same extent. Thus the implementation of a more liberal approach to development in rural areas depended to a much greater extent on local discretion, and the consequences were much more subject to local political forces than in urban areas.

During the early and mid-1980s the government took steps to reduce the scope of rural planning, but these were much more tentative and ambivalent than changes to the urban planning system, and, ultimately, under pressure from local authorities and rural interest groups, produced the opposite result. The target of reform had been the dual planning powers of district and county councils and the curtailment of the latter's strategic planning function. In 1980, the county councils lost their power to direct district councils to refuse a planning application that did not conform to the structure plan, and a series of official circulars (DOE 1980a, DOE 1984a) pressed for the structure plan system to be simplified and streamlined to guarantee ample land for development.

Finally, a consultation paper issued in 1986 by the Department of the Environment proposed the abolition of structure plans and this was followed up by a White Paper, *The future of development plans* (Secretary of State for the Environment 1989) in January 1989 that envisaged a much simplified planning system, and one that would involve greater steering from the centre. Structure plans would be replaced by regional guidance circulated by the Secretary of State after consultation with local authorities and other interested bodies, and by district development plans covering the whole of each district. Counties would issue "County Statements of Policy" that would take account of regional guidance and be only of an advisory nature.

Politically, the proposal to abolish structure plans was an attack upon the planning functions of county councils, which in England were mostly Conservative-dominated, and was resisted by them. Equally, the recurring suggestion that the counties should be abolished has elicited a strong rearguard action in their defence (Association of County Councils 1990). Against this opposition, the government could count on the support of rural district councils, which are also Conservative-dominated, and which would be accorded an unrivalled rôle in subnational planning. On the other hand, opposition to the weakening of county councils by conservationists, usually well organized and influential at the county level, would have counted for less with the Secretary of State for the Environment than the acceptance of developers.

However, a major rethink on the reform of the planning system followed the appointment of Chris Patten as Secretary of State in the autumn of 1989 (Flynn & Lowe 1992). Patten's appointment was presented as part of the "greening" of the Conservative government and was itself a defeat for the New Right in the Cabinet; the shift in allegiances it indicated eventuated in the dispatch of Mrs Thatcher herself the following year. Among Patten's early actions was the reversal of a controversial decision taken by his more right-wing predecessor Nicholas Ridley. Ridley had favoured the building of a 4,800-home, private "new settlement" by Consortium Developments at Foxley Wood in Hampshire. In reversing the decision, Patten made no mention of the outrage that Foxley Wood had caused among local residents and conservation groups, and the likely reverberations at the forthcoming Party Conference. Instead, the volte-face was justified on the grounds that planning decisions should be made at the local level by counties and districts whenever possible, indicating that structure plans would at least gain a stay of execution and perhaps a reprieve.

Confirmation of the reversal of the government's position on structure plans emerged almost a year later in the autumn of 1990. In a well publicized statement, Patten claimed: "There is now much wider agreement on the direction in which the [planning] system should move. As a result we can now look to achieving the necessary changes within the existing framework rather than by completely recasting it." (DOE 1990a)

There were to be changes, however, in the format of structure plans. Much of the detail the documents had previously contained was to disappear and counties were to concentrate only on key strategic issues, such as industrial development and the scale of new housing. While these changes were portrayed as further attempts to speed up the operation of the planning system and to lessen the burden on central government – counties will be able to adopt their structure plans after public consultation rather than having to submit them to the Secretary of State for approval – they also downgrade the documents somewhat. Structure plans must be consistent with regional guidance issued by the Secretary of State. Although county councils will normally play the lead rôle in submitting regional advice to the Secretary of State, it is expected that they will take "account of comments from government Departments, business organizations, development interests, and bodies representing agricultural and conservation interests" (DOE 1992b, para. 2.3.).

What is also apparent, though, is that the new format structure plans envisage greater scope for planning in rural areas than previously. Eight themes are outlined for inclusion in structure plans, and four of these have direct relevance for rural areas: green belts and conservation in town and country; the rural economy; mineral working and protection of mineral resources; and tourism, leisure and recreation (DOE 1992b, para. 5.9.). Two points in particular emerge from these strictures. The first is the contrast between the information and policies counties are to provide within their structure plans for rural and urban areas. The government has reiterated its determination that the planning system should restrict itself to land-use matters – that is, to say, the physical development and use of land – but is now to require policies relating to the rural *economy*, by no means a traditional planning concern.

Secondly, with proposals for district-wide development plans still in train, a major step in the extension of the planning system to the countryside is under way. Over the years this has been a gradual but accelerating process. As we argued in Chapter 3, under the 1947 Agriculture Act and the ensuing productivist period, agriculture was effectively excluded from planning control. The so-called "comprehensive" postwar planning system was in effect town planning (Hall et al. 1973). Beyond the town or village envelope, the vast bulk of rural land was not covered by development plans. It was simply designated "white land": land where there was a presumption against industrial or housing development. Only with the advent of structure plans in the late 1960s did the planning system begin to take an overview of rural land use. Even so, countryside policies tended to be sketchy and broad-brush, and the coverage of rural areas by local plans was, and still is, very limited. According to the government: "As of the middle of 1989, 60 out of 333 non-metropolitan district councils in England and Wales had local plans on deposit or adopted which fully cover their areas. Some 55 had no local plans at all, and many of the rest had local plans for only some of their areas" (Secretary of State for the Environment et al. 1990: 84).

Thus the effect of introducing development plans covering the whole of each district will be to bring all rural areas under the local development plan making process, in most cases for the first time. A survey by the District Planning Officers Society (1989) found that 60 per cent of all districts expected to have adopted district-wide development plans by the end of 1991, and 80 per cent by the end of 1992. Thus, not only are county councils to retain some of their strategic planning functions, but the content of rural planning policy has also been significantly enlarged and the extent of the coverage of rural areas by development plans is to be considerably expanded.

Indeed, the spread of district-wide development plans, a commitment to county structure plans, the revival of regional planning and the continued reliance on localized special area plans – for example, to cover an expanding town – as well as the requirement on counties to produce local plans on minerals and waste disposal have led to a proliferation of plan-making. Despite attempts to impose some consistency, there remains an absence of the kind of hierarchy whereby one plan would conform to the one above it. Instead, they simply overlay one another, although with the district councils now the main planning and development control authorities it is likely that the system will consolidate at this level. Legislation to make district-wide local plans mandatory and to streamline structure plans was introduced in the 1991 Planning and Compensation Bill. It was described by a staff member of the Council for the Protection of Rural England (CPRE) as "one of the most important pieces of environmental legislation in the past 20 years" (Burton 1991: 70). The CPRE inspired a late amendment to the Bill that sought to subordinate the long-standing presumption within statutory planning in favour of development to the principle that the plan should be paramount, or in the wording of the amendment, that

111

development control decisions "shall be made in accordance with the plan unless material conditions indicate otherwise".

Quite what effect the amendment will have on planning decisions remains to be seen, but revised planning guidance rushed out shortly before the March 1992 General Election no longer mentions a presumption in favour of development, noting instead that "applications for development should be allowed having regard to the development plan and all material considerations" (DOE 1992c, para. 5). While this might appear to give primacy to the plan, the Department of the Environment makes it clear that the planning system "fails in its function whenever it hinders or prohibits development which should reasonably have been permitted" (DOE 1992c, para. 5). Meanwhile, Sir George Young, the Conservative Planning Minister, sees the changes stemming from the 1991 Act as underlining "the importance we attach to the development plan and its rôle in the decision-making process. Decisions on planning applications should be made on a rational and consistent basis. They should not be arbitrary. In other words, the system should, as we have been saying for some time, be plan-led" (quoted in *The Planner*, 21 February 1992).

There is undoubtedly a certain ambiguity in the Conservatives' approach: an attempt to maintain flexibility in planning policy while creating ever tighter bureaucratic structures. For local planning authorities, the consequences may be twofold. First, there is likely to be greater variability between authorities in local decision-making as the differing balance of political forces leads to contrary interpretations of what counts as development that should reasonably be permitted. Secondly, there will be a greater involvement of local interests in plan formulation, and in some areas greater contestation.

Regulating the deregulated: the emergence of rural planning

The planning system has been remarkably successful in protecting agricultural land from external development. Today, about 75 per cent of the land in England and Wales is classified as agricultural, and the evidence is that less agricultural land is now being lost to development than in the past. Of course, the rate of land loss from agriculture to urban uses is crucially dependent upon the level of wider economic activity. Nevertheless, the information in Figure 5.2 would suggest that, even in periods of economic boom, successively lower rates of agricultural loss have been experienced over the postwar period.

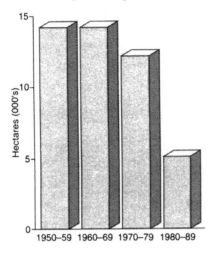

Figure 5.2 Rates of transfer of agricultural land to urban, industrial and recreation use in England, 1950–89. (*Source:* DOE 1992d. These statistics should be treated with caution as the basis upon which they were compiled changed in the 1980s, and may now be less reliable in recording year-on-year transfers.)

While agricultural incomes remained reasonably satisfactory, and the growth in land values continued, farmers, if they were interested, could look with tolerant eyes upon the planning system. Agricultural land was safeguarded from development, whereas farming operations were not subject to planning controls. This situation owed much to an agricultural fundamentalism that still pervades the rural land-use debate, with agriculture being seen as having a pre-emptive claim on the use of rural land.

The protection of agricultural land was a primary consideration in the postwar planning system. Local authorities were required to consult the Ministry of Agriculture, Farming and Fisheries (MAFF) when preparing development plans, and when considering planning applications that departed from approved plans and would lead to the development of a significant area of agricultural land. Until 1971 this was set at 2ha or more, and was then raised to 4ha. In 1978, the Advisory Council for Agriculture and Horticulture in England and Wales (1978), under the chairmanship of Sir Nigel Strutt, recommended that the threshold above which MAFF should be consulted should be returned to 2ha, "to provide stronger safeguards against the insidious erosion of the stock of agricultural land".

Just eight years later, however, faced with falling agricultural incomes and a politically unacceptable level of surpluses, the Conservative government set up an interdepartmental working party on Alternative Land Use and the Rural Economy (ALURE), involving officials from various departments with an interest in the countryside, including MAFF and the Treasury, and the Department of Trade and Industry, Department of Employment and Department of the Environment. Within the Cabinet there were rumoured to be pressures for a radical restructuring of planning in the countryside to stimulate the rural economy by allowing businesses far greater locational freedom. Such views were not unsympathetically received by the then Environment Secretary, Nicholas Ridley, an outspoken free-marketeer.

Not surprisingly, therefore, it was laissez-faire sentiments that were to the fore in a draft circular from the Department of the Environment entitled *Development involving agricultural land* (DOE 1987). This circular suggested that development proposals should only be referred to MAFF if they involved the loss of high-quality (i.e. Grades 1 and 2) agricultural land and were of 20ha or more. For land outside of Areas of Outstanding Natural Beauty (AONB), National Parks and Green Belts, the presumption that farming had first claim on the countryside was to be replaced by a view that it was of equal importance to the economic and environmental aspects of development. However, other areas of "good countryside" not covered by such designations were to be protected by local authorities from development. "Good countryside" was not defined, but clearly the implication was that not all rural land was worthy of protection and, inevitably, the draft circular provoked a storm of protest from conservationists who feared that the barriers to the proliferation of development in the countryside were being lowered.

In its final version the circular (DOE 1987) made a number of concessions to conservationist opinion. The MAFF was now also to be consulted about development on grade 3a land, and attention was directed towards the prospects of re-use of urban land. Most importantly, instead of safeguards for "good countryside" there was a promise to "protect the countryside for its own sake", rather than simply for its agricultural value. Quite what this means has never been made clear. But the apparent renewal of the government's commitment to the protection of rural land – this time for reasons other than to protect the rôle of agriculture – provides local authorities with even more scope for refusing unwanted planning applications on green field sites. Thus "the clarity of the original intention – to give less emphasis to agriculture in development decisions – appears to have been obscured" (Blunden & Curry (eds.) 1988: 97). Even so, MAFF participation in the planning system has been greatly diminished. In 1988, MAFF and the Welsh Office responded to about 1,400 consultations from local authorities for non-agricultural development, but it is anticipated that under the new rules this will drop to a maximum of 200 a year. It has been suggested, in addition, that MAFF is objecting to a smaller proportion of the development proposals referred to it (CPRE 1989).

In more general terms, ALURE altered the policy framework affecting development in the countryside, and its proposals (including encouragement for farm diversification, barn conversions and other forms of housing and light industrial development) were drawn together into a Planning Policy Guidance (PPG) note entitled *Rural enterprise and development* (DOE 1988a). Whereas conservation groups feared that the government was straying too far from established commitments to protect the countryside, agricultural opinion considered that the reforms did not go far enough to allow farmers the scope to remedy problems of overproduction and declining farm incomes. Michael Jopling, the then Minister of Agriculture, secured £25 million from the ALURE exercise to encourage farm diversification and measures to allow landowners scope to develop their land. Re-use of redundant farm buildings was encouraged and occupancy restrictions on surplus farm workers' cottages relaxed. Nonetheless, the piecemeal and opportunistic approach to reform represented by this package was insufficient to allay the chronic uncertainty and apprehension within the agricultural industry, and when Jopling announced the measures at the National Farmers' Union (NFU) annual general meeting he received an unprecedented vote of no confidence. Despite the poor reception from farmers, the government pressed on with its strategy of promoting farm diversification as a solution to agriculture's budgetary and surplus crises. The 1988 Farm Land and Rural Development Act introduced a distinction between an agricultural business and a farm business, with the former referring to the pursuit of agriculture and the latter referring to a non-agricultural business conducted by a farmer. This paved the way for MAFF to provide up to a quarter of the capital costs incurred by farmers setting up ancillary businesses under its Farm Diversification Grant Scheme. The scheme explicitly excluded "heavy industry, or light industry not normally expected in a farm environment", but specified a number of types of eligible enterprise, including the processing of farm produce, craft manufacture, agricultural machinery repair, farm shops, holiday accommodation, catering, sports and recreation facilities, livery and educational facilities relating to farming and the countryside.

The effects of declining farm support on both farm incomes and land values have increasingly preoccupied the two major agricultural interest groups, the NFU and the Country Landowners' Association (CLA). Up until the time of ALURE, the CLA directed much of its attention towards alternative ways of gaining income from the land and did not seek to challenge the planning system. Indeed, the Steele Addison Report (1984) on the integration of agricultural and environmental policy had raised the question of whether planning controls on the siting and design of farm buildings should be *introduced*. Since ALURE, there has been a shift in the CLA's outlook towards a relaxation of planning controls to allow rural landowners to diversify their activities. The Greenwell Working Party Report on the effects of the planning system on landownership was particularly damning, with its comment (1989: 12) that:

> To assist the economic regeneration of the inner cities, the government has felt compelled to suspend the operation of the planning system and introduce Simplified Planning Zones. There are few signs of an equivalent awakening to the economic needs of rural areas. The countryside is viewed as the conservation counterweight to economic revival in the cities.

Although the Working Party did not go so far as to recommend rural SPZs it did hope that a number of potentially far-reaching changes could be introduced to increase the flexibility of countryside planning. It also called for farmers and landowners to be recompensed for providing valued landscape and amenity benefits through an Environmental Land Management Scheme whereby a landowner would be able to negotiate a contract with central or local government, or a local group, on the management of all or part of a farm. The Countryside

Stewardship Scheme, introduced in 1990 by the Countryside Commission, embodied this principle.

In contrast to the CLA's demands for a liberalization of the planning system linked to positive incentives for diversification and conservation, the NFU has adopted a much more cautious approach. This in turn reflects the divisions between landed and productive capital and the distinct dilemmas and opportunities each faces in a post-productivist era. Thus the CLA has sought to facilitate owners to exploit assets through market mechanisms, whereas the NFU has clung to the agricultural support system, suitably reformed to include financial assistance for conservation and production controls, as the mainstay of its members' livelihood. It has tended to dismiss alternative land uses as having only a limited potential impact on total land use (National Framers' Union 1989/90). Nor does the NFU envisage large amounts of agricultural land being converted to other forms of permanent development, believing that not only is it politically unlikely that there will be a liberalization of planning policy to allow extensive development of agricultural land, but also that "it is doubtful if such a move would be in the interests of the farming community as a whole". This is because, first, the premium on land sold for development "is partly a result of the scarcity value created by planning controls. Widespread relaxation of these controls would substantially reduce that premium". Secondly, it is usually the landlord who benefits from development and diversification opportunities whereas the tenant farmer tends to miss out: indeed, a successful planning application for non-agricultural use could provide the grounds for a notice to quit. Thirdly, development in the countryside brings the pressures of people and traffic, and although this benefits some farmers, it causes difficulties and disruption for many others seeking to maintain their normal farming operations (National Farmers' Union 1989/90, para 29).

It seems inevitable that the agricultural community and the planning system will come into greater contact even if there are no moves to diversify farming. One source of friction arises from the intensification of agriculture and its increasingly industrial character (Scrase 1988). As yet, the planning system has demonstrated neither the collective will nor the regulatory wherewithal to catch up with these changes. Under the UCO and GDO, it still uses a definition of agriculture formulated in the 1940s that is based on the assumption that it is a primordial activity which, in harnessing natural processes, is set apart from other productive activities. Nevertheless, as agriculture becomes more intensive and scientific – under the impetus, for example, of the bio-technology revolution – so the gulf between the reality and the planning definition will widen, and pressures for the regulatory framework to be updated will increase.

The most immediately relevant factor bringing farmers into contact with the planning system is pressure for farm diversification. This, by trying to extend the economic activities that farmers carry out on their land, may well involve them in development as defined by the planning legislation. The CLA has argued that the definition of agriculture used in the planning legislation should be broadened to follow that of the Farm Land and Rural Development Act of 1988. This would allow a much wider range of farm-based activities to take place without planning permission, and would mean that a decision to diversify could be taken without reference to the local planning authority (Greenwell Working Party Report 1989). But of all the freedoms from planning control offered by the GDO, that for agriculture and forestry is already greatest. This takes two forms. First, it allows farmers complete flexibility in choosing between different methods and types of forestry and agriculture. Secondly, most building or engineering operations carried out for agricultural purposes are also classified as permitted development. But with planning law running far behind the changes in agriculture, disputes over what counts as a farming operation are likely to grow, involving greater recourse to the courts to settle matters.

Indeed, government is under pressure from both amenity and conservation interests and rural planning authorities to put a check on certain agricultural freedoms, while also seeking to create new development rights in order to facilitate farm diversification. Thus, in 1986, the government amended the GDO as it related to agricultural operations in an attempt to limit excavations on farmland to legitimate farming purposes (e.g. fish-farming) rather than for the sale of minerals. A similar move was made to limit the tipping of off-farm materials on agricultural land (Grant 1988). In 1988, further restrictions were introduced on the building of intensive livestock units. They are no longer automatically allowed to be sited within 400 metres of a non-agricultural building, for example.

Such regulations, along with others introduced under non-planning legislation – for example, limestone pavement orders, nature conservation orders and the constraints on farming change in Sites of Special Scientific Interest (SSSI), all introduced under the Wildlife and Countryside Act 1981 – indicate the government's ultimate preparedness to concede specific restrictions on farming practices in response to strong environmental and conservationist pressures. Under the European Community Directive 85/337, moreover, which came into effect in 1988, proposed afforestation schemes and certain agricultural developments – including land drainage and reclamation, the improvement of uncultivated land or semi-natural areas, intensive livestock units and salmon farming – are required to be subject to a formal environmental assessment procedure if it is anticipated that they will have a significant environmental effect.

In a consultation paper issued in May 1989, however, the government also put forward controversial proposals on the ways in which the GDO might be relaxed specifically to encourage farm diversification:

The objective of extending permitted development rights under the GDO to a range of recreational and other activities would be to ease the burden of planning controls on farmers and others who seek alternative uses for surplus agricultural land and buildings by diversifying their activities without compromising the ability of the planning system to protect the rural environment (DOE 1989b, para. 6).

If the government had carried through its proposals, which were along the lines of those put forward by the CLA, it could have heralded widespread changes in rural land use, and would have marked a significant step in the redefinition of the countryside. There were suggestions, for example, that agricultural land and farm buildings over five years old could be used for equestrian activities, educational purposes related to agriculture or the countryside, the display and sale of locally produced goods, outdoor sport or recreation, and the sale of food and drink to visiting members of the public. Realizing the sensitivity of its proposals, the government posed the question of whether or not they should apply in National Parks and AONBs. Previous liberalizations of the GDO in 1981 and 1985 had specifically exempted these areas.

Not surprisingly, such proposed safeguards failed to appease either environmentalists or local authorities. When announcing his decision on the consultation paper at the 1989 Conservative Party Conference, the Secretary of State for the Environment, Chris Patten, noted that

Many groups and individuals have responded to that

119

document. They have argued, and they have argued strongly, that these suggested relaxations would run too great a risk of damaging the countryside. Change will happen, they concede, but it must be properly controlled. I can tell the Conference today that I have been impressed by that argument. I am therefore withdrawing the proposals immediately (Conservative Party 1989).

Later in the year the government issued a draft revised version of its PPG on *Rural enterprise and development* (DOE 1988a), now entitled, significantly, *The countryside and the rural economy* (DOE 1992d). Not only did this reaffirm the government's commitment to rural planning as a device to reconcile development pressures with safeguarding the countryside, it went further to state that: "While the government has no present plans to extend planning controls to all farming activities, it is ready to introduce new closely targeted controls where this is necessary to deal with specific problems" (para 22).

Conservation groups, such as the CPRE, seized upon this suggestion to propose a variety of farming activities over which the government might introduce planning controls, arguing forcibly for controls over farm buildings in particular (Council for the Preservation of Rural England et al. 1990). The CLA, meanwhile, quickly registered its alarm and emphasized that "sustainable development in the countryside will only be possible if choice and enterprise are allowed to flourish" (Country Landowners Association Press Release, 28 February 1990).

The debate, though, was moving in favour of the conservation lobby. The Environment White Paper (Secretary of State for the Environment et al. 1990) promised discussions on the matter, and less than a month later a DOE consultation paper duly appeared on *Planning control over agricultural and forestry buildings* (DOE 1990b), which anticipated some of the measures introduced under the Planning and Compensation Act. Along the way, though, any intention to regulate farm buildings had been diluted, such that permitted development rights under the GDO were qualified but not revoked. Farmers are now required to notify the local planning authorities of their intentions to erect an agricultural building, to allow the authority an opportunity to determine whether or not its approval is needed. If the authority so decides – and it can only challenge the siting, design and appearance, and not the principle, of the development – then the farmers must submit the appropriate details for approval. Needless to say, these arrangements have pleased neither farmers nor conservationists: the former, because they are seen to be cumbersome and interfering; and the latter, because they are seen to fall well short of planning controls. Farm buildings thus join a list that includes landscape conservation orders, hedgerow protection and freedom of access to commons, where government has appeared to concede the conservationists' case only to balk at the consequences of introducing any significant restrictions on rural property rights. On such issues, ministers have been torn between wishing to appease the conservation lobby while not offending landowning interest, and the result has been stalemate.

On other matters, though, a practical resolution of these competing concerns has been sought through a geographically differentiated approach. Indeed, the government has expounded the idea of a differentiated countryside that calls for distinct approaches to environmental protection (DOE 1992d). In areas statutorily designed for their landscape or wildlife qualities, including National Parks, AONB and SSSI, policies of restraint are expected to prevail. Additional statutory planning controls apply, for example, through tighter controls over permitted development rights, and planning policies and development control decisions should "sustain or further the purposes of . . . designation" (DOE 1992d, para. 3.1.).

The most productive farm land is also given some protection, but through the normal operation of the planning system. About one-third of agricultural land in England and Wales is of grades 1, 2 and 3a, and local authorities are advised by government to give "considerable weight" to its protection "as a national resource for the future" (DOE 1992d, para. 2.5.).

For other areas of the countryside a different approach is prescribed, relying much more upon the virtues of development, on the grounds that: "Maintaining a healthy rural economy is one of the best ways of protecting and improving the countryside" (DOE 1992d, para. 1.6). Retaining as much rural land in agricultural production should no longer be a priority, and, in preparing development plans and considering planning applications, authorities should note that:

> Land of moderate or poor quality (grades 3b, 4 and 5) is the least significant in terms of the national agricultural interest [and] little weight need normally be given to the loss of such land, except in areas such as hills and uplands where particular agricultural practices themselves contribute to the quality of the environment, or to the rural economy

in some special way. (para. 2.5.)

What appeared to be a more developmental attitude to the countryside, along with encouragement of the re-use of farm buildings, whether or not agriculturally redundant, aroused fears amongst some conservationists that the government was "subordinating countryside protection to industrial and other development" (Council for the Preservation of Rural England Press Release 22 January 1992), and so seeking to turn the clock back to the ALURE period (see above). The situation, though, is quite different: with the intervening moves towards a plan-led system and the steady entrenchment of conservation interests at the local level. As the CLA has observed, a flexible planning policy is desirable "but [it] must be interpreted positively" at the local level (CLA Press Release 30 January 1992).

In retrospect, the reform of the Use Class Order in 1987 may come to be regarded as the high-water mark of liberalization of the planning system in rural areas. In a number of respects it has affected rural planning. Out-of-town land allocated for industrial development, for example, has been developed for other business or commercial uses. In addition, Home (1987: 70) has written that, "of all the use classes in the 1987 UCO, the leisure and assembly class has the greatest potential impact upon the countryside and coastal areas". The class is a combination of two previous ones, covering such activities as cinemas and skating rinks, and Home's fears for rural areas are based on two counts. The first is a failure to define or distinguish "sport", "recreation", "assembly" and "leisure" within the UCO. Consequently, the class covers a broad range of activities. The second is that the GDO offers large freedoms to such activities as camping and caravanning and amusement parks. As Home (1987: 70–1) remarks: "It appears to offer freedom for the owners of extensive facilities such as golf courses or football grounds to turn their sites into whatever recreation uses they think fit (e.g. amusement or theme parks) . . . A farmer who establishes a recreational use on part of his land such as seasonal camping can now enjoy the freedom of this class from planning control." In short, Home believes that the new class confers upon the leisure industry effectively the same freedom from planning control as farming.

The effect of such efforts to liberalize planning restrictions on the development of rural land will be to increase the access to it of non-agricultural capital. This, in itself, will raise the spectre of more local protest. Where development requires planning permission, the government has encouraged local authorities not to be hidebound by a preservationist attitude. Such advice and exhortation has been repeated in a series of planning circulars covering industrial, commercial and residential development. The government's initial attitude to industrial development in the countryside was outlined in its main planning circular for England and Wales, *Development control policy and practice* (DOE 1980a). It sought to leave to the judgement of the private sector questions to do with the demand for, and appropriate location of, industrial and commercial development. In keeping with this stance, it has pressed local authorities to accept that a much wider range of economic activities is suited to rural locations. The circular therefore urged that when "small scale commercial and industrial activities are proposed especially in existing buildings, in areas which are primarily residential or rural, permission should be granted unless there are specific and convincing objections such as intrusion into the countryside" (DOE 1980a). The introduction of new jobs into rural areas, it was argued, would "prevent loss of services and keep a viable and balanced community". The circular did, however, add an important caveat to its pro-development stance: that "the government's concern for positive attitudes and efficiency in development control does not mean that their commitment to conservation is in any way weakened".

Nevertheless, the tension between development and conservation has been a difficult one for the government to resolve. There has, perhaps, been least controversy surrounding rural economic development. Circular 16/84 on *Industrial development* (DOE 1984b) stressed the need for new enterprises in the countryside and commented that "many small-scale buildings can be fitted into rural areas" without causing unacceptable disturbance. The desirability of converting redundant buildings was reiterated, and planning authorities concerned about intrusion into the countryside were reminded, in a somewhat double-edged argument, that many light industrial uses are less noisy than some agricultural activities. Authorities were also urged to be more flexible and responsive to employment-generating proposals and not to be hamstrung by restrictive policies in development plans.

Circular 2/86 (DOE 1986) specifically addressed the needs of development by small businesses. In an apparent contradiction of Circular 22/80 (DOE 1980a), reflecting the shifting attitude within government towards the protection of rural land, local authorities were exhorted not to reject proposals "merely because planning policies restrict development in the countryside. The main object of such policies must be to protect the landscape from unacceptable new buildings, and they should not be used to prevent new uses or development whose impact on local amenity and infrastructure would be marginal." This seemed to be an attempt to moderate rural restraint policies and to limit them to questions of design and landscaping. The new PPG (DOE 1992d) takes this process one stage further in specifying that "sensitive, small-scale new development can be accommodated in *and around* many settlements" (para. 2.13, emphasis added), thus appearing to erode the concept of a village envelope.

Local authorities do seem to have responded to pressure to allow more employment uses into the countryside, but this is highly variable spatially. The relaxation of preservationist policies has tended to favour less the development of open land and more the re-use of existing buildings, including surplus farm buildings, large country houses and redundant institutional complexes (such as mental hospitals). Government-commissioned research on rural employment concluded that the planning system "had not represented a severe restraint on the development of small businesses." (JURUE 1982). The exceptions were activities thought to be too intrusive or out-of-keeping such as car-repairing, scrapyards and haulage-contracting.

There has been controversy, however, over the extent to which the greater returns on residential conversions of traditional farm buildings have precluded productive uses (Watkins & Winter 1988). It has also been claimed that the stock of re-usable farm buildings is becoming exhausted in certain regions (McLoughlin 1989). This is likely to lead to pressures to convert modern agricultural buildings that have become redundant, such as milking parlours, and to construct purpose-built units, and not just for new or incoming firms. The dispersed development of small-scale rural businesses over the past decade or so is likely to generate its own pressures for additional development that may be difficult to resist.

Far more controversy has been generated, especially in the south-east, by the government's efforts to make more housing land available. Throughout the successive Conservative administrations, government has lent heavily in favour of developers. One of the main impediments to the supply of private housing was thought to arise from an unnecessarily restrictive approach to granting developers planning permission. As Circular 9/80 on *Land for private housebuilding* (DOE 1980b) declared: "the availability of land should not be a constraint on the ability of house builders to meet the demand for home ownership". To this end, the government required local authorities to co-operate with housebuilders in identifying land with planning permission for five years ahead and to make up any shortfall. These so-called joint land availability studies were to play a key rôle in determining the implementation of planning policies, including appeal decisions. The subsequent Circular *Land for housing* (DOE 1984c) sought to ease further the supply of land by requiring all districts to have at least two years' supply of housing land immediately available for development. If they did not, there would be a presumption that planning applications for housing should be granted.

The effect of these two circulars was to give greater prescriptive force to projected housing requirements contained in structure plans, in determining applications. In having to achieve these "targets", the risk was that greater latitude had to be accorded to the developers' choice of where demand should be met, even if this involved departures from other planning policies or considerations. However, developers have not had it all their own way. Disputes over the conduct of joint land availability studies led in June 1987 to the withdrawal from some of these exercises of the House Builders' Federation, who were unwilling to accept needs-based projections, particularly in areas of high housing demand. In an attempt to revitalize land availability studies the government has issued guidance on their conduct (DOE 1992a) but has underlined its shift of attitude towards the planning system by making the housing provision policies of adopted structure and local plans the basis of discussion. In any case, legal challenges by local authorities against appeal decisions have effectively restored to them some discretion in balancing relevant material considerations even against a shortfall in housing land supply (Hooper et al. 1988). When launching a new draft of the PPG on Housing the Secretary of State noted that "We are not in the business of sacrificing environmental quality to sheer housing numbers" (Patten 1989). In its revised form, PPG3 (DOE 1992a) goes still further and withdraws the special presumption in favour of releasing land for housing by cancelling Circular 22/84 (DOE 1984c) (see above).

Ministers' attempts to relax the constraints to development posed by Green Belts, and to reduce their geographical coverage, have proved even more contentious. In the early 1980s, a draft circular to this effect aroused the ire of conservationists and backbench Tory MPs. In the end a campaign successfully orchestrated by the CPRE, the opposition of more than 60 (mainly shire) Conservative MPs and an inquiry by the House of Commons Environment Select Committee forced the government to retreat. In the approved circular (14/84) (DOE 1984d) the government emphasized the continuity of Green Belts and their rôle in assisting urban regeneration. As PPG 2 (DOE 1988b) puts it:

> The government attaches great importance to Green Belts, which have been an important element of planning policy for more than three decades. (para 1)

> The essential characteristic of Green Belts is their permanence and their protection must be maintained as far as can be seen ahead (para 7).

As we have already seen, the government encountered a similar backlash in 1987 when it attempted to encourage greater diversification of the rural economy. These struggles between ministers and backbenchers continued into 1988, as the government looked for greater release of land for housing in the southern counties and, in an unusual display of organized dissent, over 90 Tory MPs formed a group called Sane Planning to resist additional major development pressures on the countryside. With Michael Heseltine, the dissident former Cabinet Minister and ex-Environment Secretary a leading member of the group and with the increasingly public and trenchant exchanges between him and the then Environment Secretary, Nicholas Ridley, the dispute acquired a wider political significance in terms of the direction of the Conservative Party. It signalled the possible relevance to Conservative philosophy, post-Thatcher, of planning, the regulation of market forces and dirigisme generally.

The chords of discontent in the countryside that Heseltine and his allies were able to play upon so successfully were those expressed by the service class. Mainly ex-urban newcomers, with none of the political quiescence of the rural working class, they challenged the political and social leadership of farmers and landowners, gradually taking over many of the established institutions of rural society and creating new ones reflecting their own interests and particular visions of the rural community. One of the most pervasive expressions of the service class in the countryside is the growth of local amenity and conservation groups (Lowe & Goyder 1983). Once settled in their chosen town or village, its members are reluctant to see changes that might adversely affect the environmental features that first attracted them. An "Englishman's home" may be his "castle" but the space to be defined now includes the local neighbourhood and its surroundings. In an increasingly plan-led system, the articulate and well organized middle class, prevalent throughout lowland England and spreading elsewhere, are likely to play an increasingly important rôle in determining the nature of the post-productivist countryside.

Conclusion

Overall, the Conservative governments since 1979 have presented a significant challenge to the postwar planning system. From a New Right perspective, statutory planning was seen to act as a brake on market forces, and even if many of its instruments remain, there have been some radical reforms, most notably in the Labour-dominated inner-city and metropolitan areas. The creation of Enterprise Zones, Simplified Planning Zones and areas covered by Urban Development Corporations have all helped to weaken the planning system in an attempt to allow the market to prevail. Quite different pressures, though, have been at work in areas of high environmental value, including National Parks, AONB, SSSI, and Conservation Areas. Such areas have not only been largely exempted from the relaxation of planning powers introduced elsewhere but have also been given additional safeguards.

As in urban areas, the countryside has already been subject to ministers' deregulatory instincts. In this case attempts to liberalize the planning system have coincided with problems of agricultural overproduction and a wish to wean farmers off dependence on the state in efforts to open up the countryside to new forms of investment. So, for example, there is now less involvement by MAFF in the planning system to protect agricultural land from development.

Both the pressures for development and the form they take are spatially variable. So too, therefore, are the political and social processes that accompany accumulation and commoditization. Attempts to exploit rural space will in some areas provoke intense controversy and in others will not, but the focus for representations will nearly always be the planning system. Tensions between new forms of accumulation or commoditization predicated on private production norms, and concern to protect the environment based on arguments of public good, are mediated through the planning system (Marsden et al. 1991).

At a political level the intractable nature of these disputes has presented some problems for the Conservative Party since it draws support from both camps. It is still true to say that the countryside remains the heartland of Toryism, a more pragmatic and paternalistic brand of conservatism than that to the fore in the 1980s. It has, though, particularly in the more accessible countryside, been reconstituted and replenished by the activities of successive waves of middle-class groups. Attempts to liberalize rural planning inevitably produced a backlash from people concerned to protect their own environments, environments whose boundaries were spreading farther afield. Rural conservation interests and planning authorities (mostly Conservative-controlled) have strenuously resisted the government's attempts to relax planning constraints over agricultural land, and the rural planning system has been strengthened and not weakened as a result.

In general, though, there have been broad shifts in national planning policy and regulatory style that apply equally to rural areas. For example, planning is now less public-sector-led and more private-sector driven, less oriented towards community needs and more towards market demands, and more directed towards the provision and attainment of positional as opposed to collective goods.

In short, while inner city and metropolitan areas have experienced a reduction in locally accountable planning functions, the countryside has seen a general extension of the local planning system. In the absence of a comprehensive policy for agriculture, so long the major plank of rural policy, and with moves towards a plan-led system, there is likely to be a much more complex and diverse pattern of development, consumption and regulation in the countryside.

CHAPTER 6

Locality and power in the analysis of rural change

Introduction

In Chapter 2 we argued the need for a series of middle-level concepts to close the gap between theory and practice, and between global trends and local changes. The relevance of these concepts to our understanding of change was then illustrated through an historical account of rural development in the UK. This demonstrated that economic actors, the regulatory planning system and local political configurations are all important in shaping specific land development processes. The development of land is place-specific; although the forces governing it may be national or international, the outcomes are always localized.

These simple observations lead us to the consideration of complex questions. For instance, what kinds of relationships might we expect to find between economic and political actors operating at the national and transnational levels, and such actors operating locally? How do the forces of economic change interact with regulatory powers to condition local outcomes? What scope do locally based actors have to resist or significantly alter such outcomes? What, in this context, is rurality? Analysis at the local level must confront these issues if the multidimensional nature of rural change is to be satisfactorily assessed.

Here we shall examine two related concerns: first, the relationship between strategic and local processes, looking in particular at how this has been treated in the debates on "locality". Secondly, and leading on from this, we assess to extent to which the renewed interest in locality has led to a reappraisal of the relationship between the economy and polity of specific spatial areas. We argue that contemporary approaches are marked by implicit conceptions of power and agency that often undermine the theoretical and analytical claims being made. We conclude the chapter with a reformulation that begins to point the way forward.

Our overall concern in this chapter is the relationship between structure and action. How, in other words, can we use locally based analyses to examine structural processes while accommodating the uniqueness of place-specific action? We begin by examining studies of locality undertaken in the early to mid-1980s, when the overbearing structuralism of previous work was challenged through attempts to tie structural and local changes together in a non-deterministic fashion. However, what emerges is a theoretical gap between the two levels of analysis. More recent work in the locality tradition continues in this critical vein but we conclude that its revised methodologies also remain within the boundaries of structuralist argument. As we began to argue in Chapter 2, conceptions of social action remain weak and underdeveloped. In the third section we outline a further position that, we believe, offers a more satisfactory method for handling the range of economic, political and cultural processes underlying change in (rural) localities. This is drawn from recent work within the "sociology of translation". It seeks to understand how certain key actors establish power relationships by drawing upon both strategic and local resources, leading us to reconceptualize both "locality" and "rurality".

Restructuring localities: from the geological metaphor

So much has been written in the past 10 years on the term locality that its hard to imagine that much can be added. The term came into its own in the late 1970s, as spatial variations in social processes came to be recognized as an enduring characteristic of the industrialized West. "Locality" replaced two earlier terms – "community" and "region". By then, community studies had come to be characterized by functionalist methodologies and idealistic analyses that had left the concept indelibly tainted (Bell & Newby 1972; Day & Murdoch 1993). "Region", on the other hand, still retained its analytical utility but seemed to have been sidelined by the intra-regional restructuring of economic and social relations. As Duncan & Savage (1991: 156) put it:" 'locality' defined as local labour market replaced the earlier focus on 'region' precisely because it is easier to conceptualize links between industrial restructuring at the international scale and 'changes on the ground' at this level".

Locality was adopted as the preferred level of analysis, but the meaning of the term and its subsequent usage became shrouded in conceptual mystification. Gregson (1987) argued that the concept was used in a variety of contradictory ways, including locality as local variation in the unfolding of generalized social processes, and locality as case study method. Many texts had used the term in unspecified ways that left key questions unanswered. For instance, are general economic and social processes becoming more apparent at the local level? Or are local conditions becoming more influential? Or is it simply that the locality is the only arena in which the playing out of general processes can be observed?

Rather than attempting to summarize the locality debate as it currently stands, we focus here on key texts that have marked out the terrain of the locality approach. We start with Doreen Massey's *Spatial divisions of labour*, which perhaps more than any other book gave impetus to locality studies in the 1980s. Massey situated her analysis firmly within the restructuring approach (Ch. 2). She argued that, despite the increasing spatial mobility of capital, "local histories and local distinctiveness are integral to the social nature of production relations"(1984: 59). So, while we may observe an increased globalization of capital, "inside" the productive system itself lie sets of local relationships. In her conceptualization, this local distinctiveness is situated within the ambit of the spatial divisions of labour, for "just as the divisions of labour between different workers can increase productivity and thereby profit so can its divisions between regions, by enabling the different stages of production each to respond more exactly to their own specific location factors. Spatial structure, in other words, is an active element in accumulation" (1984: 74).

Local areas are therefore profoundly shaped by their rôles in the spatial division of labour and must be analyzed in terms of their current and former rôles and the relations between the two. To describe this process Massey adopted a now well known geological metaphor. She suggests (1984: 117–18) that:

the structure of local economies can be seen as a product of the combination of "layers", of the successive imposition over the years of new rounds of investment, new forms of activity . . . Spatial structures of different kinds can be viewed historically (and very schematically) as emerging in a succession in which each is superimposed upon, and combined with, the effects of the spatial structures which came before . . . So if a local economy can be analyzed as the historical product of the combination of layers of activity, those layers represent in turn the succession of rôles the local economy has played within wider national and international structures.

Although this conception of local areas as made up of sedimented layers of economic functions is simple in itself, it allows us to characterize localities in complex ways. It also marks a clear break with earlier "bottom-up" community studies. Although these characterized the unique nature of indigenous change, they failed to incorporate endogenous transformations beyond broad notions of "modernization" and "gesellschaft". In Massey's formulation, there is a continuing articulation between a locality's previous rôles and attempts to develop new ones. Each locality, having played a multitude of such rôles, must be treated as unique, while the processes that shaped, and continue to shape, its characteristics are the general processes of restructuring.

We should be clear that Massey was addressing a very specific set of concerns associated with the geography of industrial location and employment change. Her analysis seeks to avoid being economically determinist and, moreover, stresses that local areas "are not just in passive receipt of changes handed down from some higher national or international level. The vast variety of conditions already existing at the local level also affects how these processes themselves operate" (1984: 119). It is also clear from her writing that the metaphor is applicable to a wider set of social processes than simply those under consideration Thus she points out (1984: 120):

The layers of history which are sedimented over time are not just economic; there are also cultural, political and ideological strata, layers which also have their local specificities. And this aspect of the construction of "locality" further reinforces the impossibility of reading off from "a layer of investment" any automatic reverberations on the character of a particular area.

Localities, then, are the complex outcome of various economic and social layers of historical "deposits" and it is these that form the basis for succeeding rôles.

The formulation walks a knife edge between generalization and specification. It tries to integrate the structural and the contingent: "the challenge is to hold the two sides together; to understand the general underlying causes while at the same time recognizing and appreciating the importance of the specific and the unique" (1984: 300). Attempting to hold the two sides together is problematic, for it is all too easy to fall down on one side or the other. Thus, it is argued, that "what we see here are national processes in combination with, and embedded in, particular conditions producing the uniqueness of local economic and social structures" (1984: 194–5). As expressed here, the production of local uniqueness seems to be driven primarily by forces emanating from beyond the locality. In this sense, Massey's use of locality seems to span Gregson's categories, moving from locality as the manifestation of general processes to the locality as determinant of the outcome of such processes. Furthermore, locality is also regarded as an object of analysis. But in what sense is a locality an objective entity? What binds the potentially disparate elements in a particular geographical area into some kind of whole? Massey does not answer this question directly, but seems to imply that localities have some kind of economic unity. A similar position was also signposted by Urry (1984) when he argued that localities exhibit more unity than regions. It was put into practice in the Changing Urban and Regional System project which adopted the local labour market as a surrogate for locality (Cooke 1989). The problem with such an approach is that it requires an explanation of how economic boundaries equate with, or shape, political or cultural boundaries. This highlights the difficulty of establishing empirically just what the relations between the various spheres are likely to be in any given instance. The perspective fluctuates between over- and under-determination.

Warde (1985) identifies three further weaknesses in the use of the geological metaphor. First, a characterization of localities that sees them as having played a succession of rôles in the unfolding spatial divisions of labour fails to specify whether the mechanisms that generate each layer are the same or variable over time. Massey, along with others working in the restructuring vein (Urry 1981, 1984), argues that one of capital's prime motivations for seeking spatial advantages is the price and availability of labour power. Warde questions whether, historically, this was always the case. If not, he asks, then what are the "transformation rules" between the various layers? Only when these are understood can the complex relations between layers be uncovered.

Secondly, there is the problem of what Warde calls the "class combination rules". Decisions on the part of capitalist enterprises to locate in a particular locale will depend upon that place's combination of characteristics. These may include skill profiles, class conflict, and local political and cultural practices. Yet Massey's metaphor tells us little about how such characteristics become incorporated in capitalist decision-making. This points, Warde argues, to the need for a much clearer understanding of the "rules" that govern the interaction between classes in local areas.

Thirdly, Warde believes that Massey concentrates too much upon class effects while ignoring, most importantly, the reproduction of labour power. It would be hard to sustain this criticism in the light of Massey's more recent interventions (see, for instance, 1989) but his point does have important implications for the status of the locality concept. As Bowlby et al. (1986: 329) have noted in relation to gender and labour markets: "the problem is not simply that labour market boundaries are different for various groups, and especially for women and men, making them difficult to identify – it is also that social relations outside the workplace are not necessarily confined within local labour markets".

The locality, however defined, may be different for different groups of workers, distinguished most notably by gender, ethnicity and class, but also by age. So even the economic unity of particular places may be difficult to establish. The problems are compounded by relations outside the workplace which are unlikely to be confined within any discernible economic boundaries.

Although we have concentrated here on the weaknesses of the spatial divisions of labour approach, we should make clear that we have spent some time on this work because it defined the approach to the study of localities in the 1980s. It re-established the relevance of conducting locally based empirical research. It also attempted to place the spatiality of social relations at the heart of the analysis while seeking to shed any deterministic or essentialist trappings. In our view, however, it remains at core a structuralist work: what Duncan and Savage (1991: 58) term "spatialized structuralism". It promises much concerning local action and local response but ultimately comes down on the side of structural change. Nonetheless, what is clear is that Massey's *Spatial divisions of labour* placed the spatiality of social relations firmly on the social science research agenda, and more recent work has enhanced rather than diminished its contribution.

Restructuring localities:
agency, interests and the reconstitution of local space

Despite the definitional problems associated with the term locality, a fruitful debate, centred upon the relationship between the social and the spatial, has emerged within what can be broadly termed locality studies. This has run parallel to another social scientific debate: the relationship between action and structure. These two sets of concerns, though conceptually distinct, raise similar issues that at times elide. Here, we wish to pursue our earlier point that, although Massey's *Spatial divisions of labour* erred on the side of structuralism, it raised the question of agency. The point can be made in the following terms. What rôles do local agents play in the reproduction of larger structures? By posing the question thus, the concern for spatiality will not obscure the problem of "agency", as it is our intention to show that spatiality and agency are linked issues. One can only be satisfactorily addressed, at the very least, by taking note of the other (Long 1990).

The difficulties associated with characterizing localities have focused attention on the construction of local spaces within unfolding general economic and social processes. This raises the question of what particular elements of economic and social life, if any, are localized and why. These are questions that Cox & Mair (1991) characterize as the "scale division of labour", which they present as a corollary to the spatial division of labour. Cox & Mair argue that socio-spatial relations are localized for three main reasons. First, certain activities are necessarily constrained locally. For example, the attempt to minimize socially necessary labour time implies the need to reduce movement and the development of spatially constrained linkage patterns. Put another way, the reproduction of labour power entails some proximity between work and home. Secondly, production incorporates a degree of immobility, for instance in infrastructural provision or localized knowledge bases. Thirdly, these tendencies towards constraint and immobility are reinforced by the uneven nature of capitalist development. Geographical restructuring implies "a constant threat of devaluation of capital, labour power, and the state . . . devaluation is place specific" (Cox & Mair 1991: 199). The incorporation of spatial structures within the processes of production and reproduction thus gives rise to different spatial forms, and "hence it may be useful to add to the idea of a spatial division of labour . . . a scale division of labour" (ibid.). Whereas the spatial division of labour refers to the division of activities between separate territories, the scale division of labour refers to "the division of activities between the different levels of the hierarchy of spatial scales, the territories composing it therefore being nested" (ibid.: 200).

There are no doubt a multitude of different scales within the division of labour, some of which we could characterize as local and others as global. Clearly, the three points above refer to localizing tendencies within the processes of production and reproduction. However defined, locality represents a scale around which some degree of concreteness can be detected. Cox & Mair are keen to point out that they are not arguing for the existence of "some perfectly coherent locality" and see the locality as "embedded in scale and spatial divisions of labour, which means that each local actor is also linked, whether directly or indirectly, to actors outside the locality" (201). These links will vary between actors and they are likely to be connected to the "exterior" at different scales. So elements within the locality will be embedded within the scale and spatial divisions of labour in diverse ways.

This would seem to raise problems for any conception of locality as unified social space. Yet Cox & Mair proceed to argue that "despite the complexity of the scale division of labour, and the real fuzziness of boundaries, geographically defined social structures at the local scale, identifiable by locals themselves, do emerge" (202). They recognize that the meaning and shape of locality will vary between actors located at the different scales. They then seek to take the argument further, suggesting that in some sense the locality can be conceived of as an "agent" (204). But for locality to become agent requires two moments: first, mobilization must be defined locally and involve members of the local social structure; and, secondly, it must lead to the creation of emergent powers, over and above those represented by the sum of the actors involved. Examples might include local business coalitions that "suspend" internal divisions (associated with class or race for example) in order to attract outside investment or mobilize resources for local economic development. Such coalitions "speak for" places.

While Cox & Mair alert us to further complexity in the relationship between the local and the structural, they do not succeed completely in "closing the gap" between the two. This derives partly from their unproblematic view of actors and social action. They counterpose local action with that which takes place at other spatial scales, as if somehow these actions can be separated. This then leads them to see the locality as agent. In our view this is a misuse of the term agent. The locality does not act, the agents "within it" do; and seldom in unison. To conceptualize locality as an agent is a distraction. It shifts our attention away from Cox & Mair's earlier and important point that the meaning and shape of locality arises from the interaction of social actors *embedded* at different spatial scales. The point has recently been recognized by Massey (1991: 28):

what gives a place its specificity is not some long internal-ized history but the fact that it is constructed out of a particular constellation of social relations, meeting and weaving together at a particular locus . . . It is, indeed, a meeting place. Instead then of thinking of places as areas with particular boundaries around, they can be imagined as articulated moments in networks of social relations and understandings, but where a large proportion of these relations, experiences and understandings are constructed on a far wider scale than what we happen to define for that moment as the place itself, whether that be a street, or a region or even a continent.

The challenge is to specify much more thoroughly the links between actors, to examine what Massey, in her 1991 article, terms "the power geometry" (25) of the relationships that weave through space. Only then can we understand the terms and outcomes of their "meetings" in "places". The task may be approached through a discussion of "interests".

In conceptualizing the locality as agent, Cox & Mair run the risk of ascribing interests to places or to all the social agents within such places, purely on the basis of their geographical co-existence. This is problematic, for place as a geographical entity can have no social interest. This is clearly not Cox & Mair's point. As they argue (1991: 208): "locality as agent defined as the sum of individual local actors is certainly problematic. But locality as agent defined as local alliance attempting to create and realise new powers to intervene in processes of geographical restructur-ing has now become a vital element of those very restructuring processes".

Local alliance refers to a variety of agents coming together in order to represent the place, to pursue their conceptions of what the interests of the place might be. This is very different, however, from some conception of the locality as an actor with interests. As Urry (1990) has pointed out, the notion of the interests of a locality is problematic. First, as we have already noted, localities comprise a variety of social actors operating at different scales who will identify with a territory to different degrees. Secondly, territorial identifications will not necessarily coincide (the "boundary problem"). Thirdly, separate social actors will have distinct interests in the place itself, depending on their position within the spatial and scale divisions of labour. As Urry says (1990: 189):

> Different social groups have different stakes in a place, and their interests vary from the more obviously material (which itself varies from the straightforwardly 'economic' to that of ontological security) to the cultural and aesthetic. Furthermore, some social groups will possess superior sets of resources and this may have the result that their conceptions of the interests of the locality become dominant.

From the earlier concern with the unfolding of spatial structures we thus move towards an analysis of how locally based actors insert themselves into the processes of restructuring. This is being undertaken in the full awareness that local actors are tied to external actors in a variety of ways and at a variety of scales. To examine these relationships, and to understand how certain conceptions of interest predominate, thereby requires us consistently and clearly to define "actors", "interests" and how these relate to social action and spatial change.

According to Hindess (1986a: 115), an actor "is a locus of decision and action, where the action is in some sense a consequence of the actors' decisions . . . reference to an actor always involves some reference to definite means of reaching and formulating decisions, definite means of action, and some links between the two".

A locality therefore cannot act; to say so is merely a shorthand way of saying that certain actors claim to act in the interests of the locality. These claims may strive to be legitimate representations of the locality's interests, but the locality can have no interest independent of particular social actors making such representations. Hindess (1986b: 118) sees interest as inextricably bound up with action:

> If the concept of interests is to play any part in the analysis of action it can only be because interests are thought to relate to the decisions of particular actors, and therefore to their actions. Actors *formulate* decisions and act on some of them. The concept of interest refers to some of the reasons

that may come in to the process of formulating a decision.

Action and interest are closely bound together in this conception that, furthermore, directs our attention to the conditions under which interests are constructed: "to say that interests are formulated is to insist on a further set of questions concerning the conceptual or discursive conditions necessary for certain reasons to be formulated at all" (Hindess 1986b:119).

Hindess is arguing here against any notion of "real" interest, pre-given by an actor's position in the social structure (or, for that matter, the locality). Such real interests can have no social effects unless they are recognized and acted upon by those actors to whom they refer. They therefore have to be formulated under certain conditions. As we have seen in our discussion of rural property rights (Ch. 4), interests are not detached from social structure for "the decisions [actors] formulate and the reasons that enter into those decisions depend on the discursive means" (Hindess 1986b: 121), and will be partly governed by the possibilities for action that are seen to be open to the actor. These are related to the social conditions under which the assessment is being made, and include other actors pursuing courses of action based upon alternative assessments, attempting to impose their conceptions of their interests upon others.

It is in this context that we can see the relevance of locality as a "meeting place", constructed out of a constellation of social relations. These social relations comprise sets of actors somehow tied together at different scales attempting to pursue their conceptions of their interests. The locality itself can be conceptualized as comprising layers of outcomes, to return to the geological metaphor, as actors pursue their perceived interests in competition with others. But how are actors tied into social relations? And how can certain sets of actors impose their interests upon others? It would seem that it is only by addressing the issue of power that the relationship between social and spatial change can be fully understood. Out of this understanding we can begin to grasp how locality, and of course rurality, become invested with particular social meanings. We thus need to focus upon why actors adopt particular courses of action, or, more specifically, the "rules" and resources that condition such an adoption, and on the means they employ in the attempt to reduce the amount of discretion enjoyed by other actors in the competition to impose interests. Out of this *competition* come the local and the rural.

"Meetings" in places: actors and networks of power

The argument presented so far in this chapter has brought us to a conception of place as a meeting point where sets of social relations intersect. At this point, new sets of relationships may come into being as the complex of social actors jostle for superiority, as actors formulate interests and as they attempt to impose them on others. Whether this jostling takes economic, political or cultural form does not, at the moment, concern us; power manifests itself in many different ways. What concerns us is how these sets of relationships shape the "local" and the "rural". Both are clearly an amalgam of the various social spheres. Furthermore, by adopting an approach that takes actors and their interests as its starting point, we recognize the relational character of interest formulation but do not assume structurally determined explanations of social action. This "holistic" approach to the analysis of power relations, explicitly links knowledge (the discursive capability to formulate interests), social action (the opportunity to act on such formulations) and "materiality" (the distribution of economic resources that facilitates certain courses of action).

Clearly, those who possess superior sets of resources (both cultural and material) are able to act more easily upon their formulations than those who do not. In the rural context, as we saw in Chapter 2, a key resource is property rights, which permit access to, and control over, rural land. Institutions also often allow privileged access to key resources: state institutions have law-making legitimacy and regulatory powers, and multinational corporations have huge capital stocks that can be readily switched from locale to locale. Greater resources often allow greater freedom of action, and even the means to acquire more such resources. Resource levels are therefore linked to, and facilitate, certain sets of power relations. However, it would be a mistake to assume that the processes of resource acquisition and domination necessarily go hand in hand. To show how fragile power relationships might be, we need to examine more closely how they come to be constructed.

A series of methodological issues is associated with an analysis of power relationships cast in these terms. An example of how we might proceed is provided by work in the "sociology of translation". This work not only addresses how empirical enquiry should be conducted but also how some actors succeed in imposing their constructions on particular issues and places while others fail.

According to Callon et al. (1985: 10), the rôle of analysis is to study the creation of "categories and linkages, and examine the way in which some are successfully imposed while others are not". This entails describing "without fear or favour , what it is that actors do" (5). Explaining the operation of power in these terms requires sensitivity to the kinds of explanations offered by both the actors under observation but also by the observer. To fulfil the obligation of describing what it is that actors do, Callon suggests three methodological principles. The first is "agnosticism", which entails impartiality between the actors engaged in a particular social conflict. The analyst "refrains from judging the way in which the actors analyze the society which surrounds them. No point of view is privileged and no interpretation is censored. The observer does not fix the identity of the implicated actors if this identity is still being negotiated" (Callon 1986: 200).

The second principle is "generalized symmetry" which stipulates that all the conflicting viewpoints must be explained in the same terms. The third principle is that the observer "must follow the actors in order to identify the manner in which they define and associate the different elements by which they build and explain their world, whether it be social or natural" (Callon 1986: 201). In other words, it is accepted that actors build worlds in which aspects of the social and natural are jumbled up, but come into a clear relationship with each other in the context of specific issues.

To exemplify how this method works it is worth reviewing a case study presented by Callon that shows how networks of social relationships come into being, and how power both binds the network and determines the extent to which it effects outcomes. He shows how one group of actors is able to get other actors to comply with its position. He gives an instance of how a power relationship comes into being, how it is sustained, and the range of its consequences. The study is locally based and illustrates how social relations meeting in one place lead to new sets of relationships, and the terms upon which these are constructed.

The case study is concerned with an attempt by three scientists to recruit other actors into a network of relationships that would facilitate the dissemination of "their" scientific knowledge amongst the scallop fishermen of St Brieuc Bay in northern France. The construction of this network involved five different stages ("moments of translation") "during which the identity of actors, the possibility of interaction and the margins of man-oeuvre are negotiated and delimited" (Callon 1986: 203).

The construction of the network was initiated by the scientists who, after a visit to Japan to study scallop farming, sought to present "their" knowledge as the solution to the problem of the depletion of scallop stocks in St Brieuc Bay. By informing other actors of the problem, they attempted to establish, as they saw it, their knowledge as an "obligatory passage point" (Callon 1986: 204). That is, they tried to define the problem in such a way that their knowledge was indispensable to its solution. In this moment, termed "problematization", the scientists tried to bring other actors into play on their terms and in so doing to create a stable network of actors that would, when necessary, support their position.

It is worth noting here that Callon extended the concept of actor to include both social *and* natural entities, entities being defined as any distinguishable feature in an actor's world that become identified during the process of translation. Some identities may represent obstacles to the actor's goal, while others may act as resources, but there is always a two-way working relationship between actors and entities that may well change as a result of technological advance, or political or social change. Callon is able to adopt this perspective on the social and natural because he is concerned with the *scientists'* definitions of a world that focuses on a "natural science" problem (the scallop stocks), but one that is socially constructed. In his case study, it is the scientists who initiate the translation process and it is therefore from within *their* processes that the actors and entities are identified and mobilized. He identifies the scallops, the fishermen and scientific colleagues as actors. We would contend, however, that only actors with the facilities to make decisions and act upon them could initiate and sustain such a process. The scientists attempt to represent the identity of the scallops, and clearly the scallops can (and do – see below) act in ways that undermine this representation. But could the scallops represent the scientists? The impossibility of this leads us to the conclusion that actors must be regarded as a locus of decision *and* action (a similar point is made by Collins & Yearley 1992). However, we would agree with Callon, who draws upon Touraine (1974), that actors do not exist outside the relationships in which they are enmeshed: identity runs in parallel with these relationships. The key point here, however, is that actors are bound together in networks that are made up not just of the actors themselves but also of natural and artefactual entities (Callon 1991, Latour 1991). At this initial point in the translation process, the actors had been incorporated within the network, but the strength of the relationships had not yet been tested. Callon is operationalizing Hindess's insistence that actors come to define themselves and their interests in the context of relationships with others, in the course of particular social struggles. The actors' identities and interests are not, therefore, pre-given, either by the social structure or by some set of fundamental internal attributes, but come to be defined as they weave, or are woven into, particular networks.

Callon's second moment of translation is termed "interessement". This is the stage at which the lead actors seek to consolidate their network by persuading other actors that their position *is* the correct one. In the case study, the scientists attempt to construct a network comprising the scallops, the fishermen and their own colleagues in order to attain their goal. In the process, they seek to define the identities and interests of those whom they wish to enlist as their allies, while also attempting to insert themselves into the competing sets of relationships that exist among other actors. "Interessement", therefore, consolidates the actor network through the enlistment of allies and by seeking to undermine competing associations and alliances.

The third moment is "enrolment". This takes the process a stage further by *stipulating* a set of relationships that operationalize the network. In many cases enrolment only comes about as a result of complex negotiations between actors over how their identities are to be fixed within the network. Once again, the attempt to create alliances can fail at this point and a wide variety of strategies may be employed by lead actors to ascribe particular rôles to other actors in the network.

The fourth moment is "mobilization". This extends our understanding of the network itself. Mobilization refers to the methods used to ensure that the representations of interest made by the lead actors are fixed, understood throughout the network, and accepted as legitimate by those who are ostensibly being represented. Representation is thus an issue in all network relationships:

> Properly speaking, it is not the scientific community which is convinced but a few colleagues who read the publications and attend the conference. It is not the fishermen but their official representatives who give the green light to the

experiments and support the restocking of the Bay. In both cases, a few individuals have been interested in the name of the masses they represent (or claim to represent)" (Callon 1986: 214–15).

The network is therefore composed of representatives. Its strength depends not only on the relationships between the representatives but also on the legitimacy of their representations. Their claims must be adhered to by those they claim to represent or the network will fall apart. In this case the *scientists* defined what the interests of the various actors were and then spoke on their behalf. At this moment in the translation process the entities were mobilized to legitimate the scientists as spokespersons for the network.

This, then, is the process of translation. It is a process whereby various disparate social entities and actors are brought together within a network, for which a strategically placed representative speaks. It is through this process that actors or entities gain identity and interest. These are not pre-given but are evoked as the links in the chain are forged. Once established, however, the maintenance of these links must be actively pursued. In Callon's study the final act is "dissidence": the scallops and the fishermen betray their representatives; they do not conform to the representations made on their behalf. The hold of the three scientists on them is lost; new spokespersons come to the fore and deny the representativeness of the earlier ones. These new representatives are able to challenge successfully the legitimacy of previous representations by pointing to their inability to account for the behaviour of those they represented.

Callon (1986: 224) summarizes the method of analysis in the following way:

Understanding what sociologists generally call power relationships means describing the way in which actors are defined, associated and simultaneously obliged to remain faithful to their alliances. The repertoire of translation . . . permits an exploration of how a few obtain the right to express and to represent the many silent actors of the social and natural worlds they have mobilized.

This approach indicates how we might "get inside" the construction of social relationships, by following actors as they formulate and pursue their interests. It allows us to understand how certain actors (or actor networks) are able to impose change over the interests of others. Such actions have social and physical outcomes. If we conceptualize the "local" and the "rural" as combining both social and physical elements, then we can begin to understand how networks of actors, which are both local and non-local, rural and non-rural, meeting at certain spatial points, come to represent, or speak for, particular places. They do this by mobilizing natural and social entities in ways that ensure their interests correspond to those of the "dominant" actor network.

It is important to treat Callon's case study as an insightful exemplar. It is difficult to know to what degree his particular moments of translation are specific to his research; and neither would we accept, other than as an ideal type, his sequence of moments. In practice, they almost certainly merge into each other, and any pre-ordained sequence would introduce a rigidity that is contrary to the whole approach. Moreover, rigid sequences have rarely been sustained, following empirical enquiry, by other models of social process, including those claiming to describe decision-making, the adoption of innovations or even land development (Ch. 7).

Nonetheless, we have examined this "actor-centred" approach at some length because we believe it provides a useful corrective to the earlier concentration on structuralist analysis. What we are seeking to do is to "follow actors" as they build their worlds, as they forge links with others, and as they attempt to "colonize" the worlds of others (for an example of this type of analysis in relation to environmental control and agriculture see Clark & Lowe 1992). The outcome of this process is what we commonly refer to as power relations. These relations may be "local" or "global", and they may be strong or weak, but the consequences of these relations (commoditization, preservation, etc.) will determine what is meant by the "local" and the "rural" both culturally and materially.

While this methodology allows analysis of the locality as a "meeting place", we need also to take account of Cox & Mair's reference to the scales of social processes. Local actors are linked to actors "outside" the locality, in ways indicated by Callon. Scale refers to distance, to the attempt by external actors to enrol local actors within particular networks of control. How, then, is the local represented within such networks? What means are made available to local actors by their incorporation within such networks to allow representations of the local to be made?

The question of scale can be posed in another way: what links local actors to non-local actors and how do non-local actors effect change at a distance? Latour (1987: 223) believes that "the question is rather simple; how to act at a distance on unfamiliar events, places and people? Answer; by somehow bringing home these events, places and people". This can be achieved by three means:

"(a) render them mobile so they can be brought back: (b) keep them stable so that they can be moved back and forth without additional distortion, corruption or decay, and (c) [make them] combinable so that whatever stuff they are made of, they can be cumulated, aggregated, or shuffled like a pack of cards" (ibid.).

Through such translation processes it is possible to do things in one place (e.g. the centre) that dominate another place (e.g. the periphery). So the term local has a double meaning: first, it refers to the aggregated practices of locally situated actors; secondly, it refers to the incorporation of local actors within various economic, political and social networks in which the local is "brought back" to the centre. For Latour (1982: 232), this raises two further problems; "what is done in the centres . . . that gives a definite edge to those who reside there? . . . and what is to be done to maintain the networks in existence, so that the advantages gained in the centres have some bearing on what happens at a distance?"

Latour cites the example of the census to show how this process is commonly undertaken. The census forms are distributed and collected. They allow all the surveyed households to be brought back to the centre of calculation. However, the number of forms and the scale of the information they hold is still unmanageable. They have to be reduced further to tables, graphs and summaries. Eventually, all the households will be translated into manageable statistical categories; they will have been through several "moments" of translation and have been combined through a common form of calculation that allows elements to be stabilized and transported from the periphery to the centre. The forms of calculation that result can then "stand for" or represent the surveyed households. In the form of the census a legitimate representation of the local can be produced by the centre. This legitimacy derives from the incorporation of the participating households within a common form of calculation, but depends also on the households' acceptance of the centre's interpretation of the material.

Inside the networks, processes of representation allow the mobilization of the local to the centre. Within economic institutions, for example, accountancy serves to lock local branches of a firm into the centres of calculation. Within state institutions, bureaucratic rules and standardized procedures tie local sites to the centres. There are a host of means available to the centre as its agents try to "mobilize" the periphery. Local actors often find they have only limited means of representation within such networks. By submitting to these common forms of representation, their scope for promoting local specificity is significantly reduced. Local actors can be caught within competing representations of their identities whereby, on the one hand, localizing tendencies, such as constraint and immobility, allow local identities to be constructed while, on the other, distant actors attempt to enrol them in standardized, mobile, procedures of translation. To suggest that current tendencies are largely centralizing may be to underestimate the extent to which power is often diffused within economic and bureaucratic structures. It seems perfectly reasonable to argue that in a post-Fordist world, decision-making will be more diffused within structures and between spaces. Whatever else, there must always be a two-way relationship between the centre and the local, and this relationship not only varies between networks but also within a network over time.

The construction of networks and the ability of such networks to "act at a distance" is what ties the local to the global. By examining the connections in this way we are able to specify the exact means by which the local is represented within the network. But what elements are combined and how are they mobilized? What forms of calculation are used within the network to carry the local to the centre? We are not concerned here with the simple unfolding of social structures through space but the means whereby networks of actors construct space through social practices using certain forms of calculation and representation. These forms, as well as the social practices themselves, must be examined if we are to understand the meaning of locality.

Analyzing local rural change: conceptual and methodological issues

We began by examining the spatial divisions of labour approach and found that, although it took the concept of locality seriously, it did so within a framework that privileged the structural determinants of local change. Nonetheless, the approach focused our attention upon the institutional relationships that govern the reproduction of spatial structures. What was lacking was a sophisticated understanding of the relationship between locally based action and that deriving from outside the locality. Cox & Mair introduced the scale division of labour as an addendum to the spatial division of labour. This increased our comprehension of the complex relationship between locally based actors and external actors, and led us to specify much more thoroughly how sets of social relations come into being, goals are formulated and attempts made to impose them in competition with others. We are now forced to confront the processes of representation:

○ Which actors attempt to represent the locality?
○ How do they recruit others to these representations?
○ What sort of locality is being represented?
○ How does this compare with competing representations?
○ Do the entities that are being represented adhere to, or betray, these representations?

The conception of locality as a "meeting place" for networks of power relations forces us to ask such questions in the course of locally based research.

We have so far stressed locality rather than rurality, although it should be clear that the reproduction of both entails the same kinds of social processes. In the rest of this section we shall concentrate more explicitly upon the rural. Where the locality has been conceptualized as a meeting point for sets of social relations, the rural implies a more restricted set of meetings, those associated with a distinctive type of social space. How rural space is distinguished, "marked off" from, say, urban space, can be examined by reference to the representational activities of an actor network that is quite familiar to us: that of rural sociology.

The means by which academic rural sociologists make their representations are well known. They include the combination of social and natural elements in texts, presented as books, articles, conference papers and so on. Also well known are the methods that allow the rural to be captured, mobilized within the network and transported to the centre: these include questionnaires, interview transcripts, field notes, statistics and the other methods of translation. They allow the rural to be "simplified, punctualized and summarized" (Latour 1987: 241). What kind of rural representation does this process give rise to?

Mormont (1990) has recently presented an analysis of the relationship of rural sociology to its subject matter. He traces the rôle of rural sociology in undermining or legitimizing certain dominant representations of rurality in Belgium, most notably those associated with agricultural modernization and the Catholic Church. The term rural, Mormont argues, was primarily descriptive, and applied to those areas that lay outside the systems of industrial development. "It was taken for granted that its subject matter was the least developed regions and least integrated areas" (1990:28). The emergence of rural sociology in the 1920s and 1930s, says Mormont (1990: 26):

marked a shift in the mode of legitimation of the category "rural" and of rural movements, from the religious to the scientific. The developing rural sociology abandoned the moralizing attitude formerly predominant. It considered the rural world to be a social world or a form of civilization – a rural civilization – differing from urban civilization in its distinctive values and social organization.

As in other countries, rural sociology in Belgium was inspired by American functionalism, which represented the rural as being made up of small-scale communities where personal relationships constituted the essence of social life (unlike in the city where collectivities were seen as standing between the individual and society); cultural characteristics were rooted in tradition; and local institutions and economic cohesion gave the localities their identities. Changes in the object of study in the following decades required shifting representations of that object. What developed, Mormont argues (1990: 28), "was not so much the idea that the rural had a specific function in the social field (only agriculture had such a function, as a food-producer) as the idea that social progress implied a gradual integration of rural regions and populations into economic and industrial development". Modernization allowed for a reconceptualization of the rural around the perceived threat to its continued existence. However, Mormont argues that this reconceptualization has been overtaken by events. Increased social mobility has undermined local autonomy; economic cohesion has been lost as more and more activities have become footloose; and new uses of rural space mean that networks comprised of local agents no longer redefine the locality. These uses may be neither rural nor local. The net effect of such changes is the creation:

of a new type of locality, which in any given area, is the result of the interaction of the various forces which operate, from a variety of fields, to confer value on that space. Thus in future, a local space will have to be understood, not in terms of its constituent elements but in terms of the possible combinations of externally determined forces able to confer value on it" (Mormont 1990:32).

As we have demonstrated in earlier chapters, each rural space now has a variety of uses or values imposed upon it. These priorities are the result of negotiation and struggle between the actors involved. Where the uses are subject to exchange values, commoditization is the result, but the process may also include actors attempting to value use according to other criteria such as those related to "the environment for its own sake" or aesthetic beauty. There is, therefore, Mormont argues (190: 34) "no longer one single space, but a multiplicity of social spaces for one and the same geographical area, each of them having its own logic, its own institutions, as well as its own network of actors . . . which are specific and not local".

This leads to a call for a reconstituted rural sociology whose subject "may be defined as the set of processes through which agents construct a vision of the rural suited to their circumstances, define themselves in relation to prevailing social cleavages, and thereby find identity, and through identity, make common cause" (41).

In response to these perceived changes it is noticeable that Mormont proposes a representation of rural space that has much in common with our previous conception of local space. Like Cox & Mair, Mormont sees the rural as providing the basis for actors to "make common cause", i.e. as a representation made in the assessment and pursuance of spatial strategies.

Through a combination of changes in the object of study – namely, the spaces deemed to be rural, and the assessment of them within rural sociology – new conceptions of rurality have arisen. Rural sociologists, no less than agricultural policy-makers, farming interest groups or amenity associations, promote representations of rural space in competition with others. These representations arise from the procedures of translation that permit the object (the rural) to be constructed within the network of interested actors. However, this construction is not detached from the object; as many elements as possible are retained, but they must be made mobile and brought back to the centre. Here the elements are combined in the process of re-representation that constitutes the sociological output. The success of the translation process depends upon adherence to these representations of the transported elements. The decline of community studies or modernization theory can be ascribed to a breakdown in the process of representation whereby the object (the rural) "betrayed" its representation in sociological texts. Whether the betrayal is a function of changes in what rural sociologists want to represent, or to changes in the relationship between rural sociology and other disciplines, or is a function of change in rural areas themselves, is more difficult to establish. Nevertheless it provides the basis for the next "round" of representational activity.

Furthermore these representations have "real" effects. It is particularly relevant to consider the operation of the planning system in the UK and the case of a development of rural land. The potential developer makes an assessment of his/her interest, draws up a set of plans representing the proposed development and submits a planning application. These are lodged with the planning office, where they are open to public inspection. Representatives of a local amenity group see the plans and decide to oppose the development. They set about persuading local residents to write letters of opposition to the planning department in which certain statements about the undesirability of the development for the character of the neighbourhood are made. These representations are also lodged in the planning file. Before the application is due to go before the planning committee a planning officer takes the file and writes a report that summarizes its contents for the members of the committee. This report also includes the planning officer's recommendation, reached after taking into consideration a variety of other representations, i.e. central government guidelines and directives, local policies embodied in structure and local plans, previous decisions etc. The committee then makes its decision either to approve or turn down the application. They will be influenced by the planning officer's recommendation and may simply follow it, or they may be swayed by local political pressure, personal contacts with the developer, and so on. In this process several moments of translation can be discerned. There is the initial translation of the development plan into an application; there is the collection by the amenity group of residents' representations and the encapsulation within these representations of certain elements of the locality and rurality; there is the translation of all these into the planning officer's summary, an "obligatory passage point" through which all representations must pass; and lastly, the decision. The result will either be a development or its prevention. Either way the material shape of the rural locality will have been altered or confirmed. Furthermore, the outcome will provide one more precondition for the next round of decisions.

Economic, political and social actors are consistently making such representations. By doing so, they are attempting to recruit allies and fix the identities of others in the pursuit of their interests. They attempt to forge networks and alliances, to procure resources, and impose their representations upon others. This is a routine process to be found in all areas of social life, from the most mundane to the most prestigious. The social and natural world that surrounds us, in which we live, can be seen as the aggregation of such practices.

Conclusion

The conception of social space presented here stresses the multiple uses to which a single space can be put and points to the localized effects of social actors operating over a variety of distances. This allows us to focus upon the methods adopted by such actors in formulating and seeking to achieve their objectives. It allows us to analyze the means by which interests and objectives are constructed, represented and come into effect. We have argued that locality and rurality are the representations of the outcomes of past practices within networks, and have directed attention to the power configurations that may have resulted from these previous practices. The fact that past actions provide the "standing conditions" for present and future actions helps us to understand how these conditions are skewed to facilitate the success of certain representations over others.

The standing conditions consist most notably of rules and resources. The rules will embody past representations and past struggles to fix interests in particular "passage points". But such rules are unlikely to be absolute in their effects; there will be discretion in their interpretation and implementation. Their reproduction may be the subject of struggle. The distribution of resources will also be conditioned by past practices and will allow certain actors privileged access to power points, or allow them the means to establish new ones. This distribution is by no means fixed; alliances have to be forged and maintained, representatives have to be marshalled and kept in line, and representations have to be legitimized and acted upon. The whole process is precarious, with the participants constantly striving for stability and certainty.

It will be obvious that any achievement of stability can be disrupted by innovation, something we have not examined here.

The production of new technologies, for instance, can redistribute resources among actors. Thus "domination is never eternal, never utterly set in time and space: it will invariably be subject to processes of innovation that may as readily subvert as reproduce its functioning" (Clegg 1989: 215-16). New technologies facilitate the faster and easier mobilization and recombination of entities within networks. Furthermore, "if it is true that our technological systems are undergoing significant paradigm changes [as the post-Fordist, flexible specialization, literature leads us to believe], then the kinds of persons or characters that are produced by and which sustain our polities and economies may also undergo significant change" (Barnes 1991: 897). Hence, new alliances, social formations and identities may come into being. We believe such changes can be captured most accurately by the geometrical model of power; where the local and the rural are considered as the *outcomes* of complex power relationships "meeting in places".

As we shall outline in more detail in Chapter 7, we characterize the methodological approach that derives from our discussion here as "action in context". We are concerned to follow Callon's maxim – "describe without fear or favour, what it is that actors do". This description must not prejudge actors' identities or give different types of explanation for different actions: it must literally follow the actors as they use resources, construct their identities, interests and strategies, and "attempt to impose worlds upon one another" (Callon et al. 1985: 228).

CHAPTER 7

Researching the rural land-development process
Key conceptual and methodological issues

Introduction

Our interest in the land development process springs from a concern for understanding the patterns of change in the rural arena and our need to adopt a vantage point from which to observe them. In this sense, the land-development process acts as an observational "window". Whereas other facets of restructuring, such as the labour market, could serve a similar purpose, the land-development process provides a perspective that allows the physical transformation of given places to be analyzed in terms that tie this firmly to broader patterns of economic, social and political change. From our standpoint, therefore, conceptions of the land-development process that attempt to uncover the interactions between the internal dynamic of the process and the context in which it unfolds meet the methodological position outlined in Chapter 6. But what do we mean by land development? In physical terms, land development includes changes in land management as well as land use. Each contributes to land development, although a distinction of degree and of kind can often be drawn between changes to management practice and changes to use. For example, a series of cumulative shifts in management might add up to development, as in the case of grassland improvement through reseeding, increased fertilizer applications and the build-up of stocking densities. The essential feature of land development is, however, normally a change in land use to yield either additional or new sources of income. It often involves the investment of new capital, but it may not, as in the introduction of commercial shooting to an existing woodland, or even the planned withdrawal of private capital as in the case of the current, publicly funded programme of agricultural extensification.

In spite of the frequent reallocation of capital arising from land development, an assessment of the process can rarely be reduced to an internal financial judgement based on the gains and losses realized by those who hold the beneficial rights to the property. This is not to suggest that those actors with a material interest in a development – landowners, developers, estate agents and so on – are insensitive to price. There is clear evidence to the contrary, the timing, nature and outcome of the process are all influenced by changing market and fiscal conditions and the assessment of these by the actors concerned (Goodchild & Munton 1985). The process also depends in part on the degree to which particular actors retain a beneficial interest in the property as the development process unfolds.

In other words, although capital is frequently a crucial input, and income sources may change in type and quantity as a result, more fundamentally the process changes the social and political relations surrounding particular pieces of land. Specifically, most cases of land development lead to a redefinition or redistribution of property rights, each outcome contributing to a constantly changing local setting within which the next round of contestation takes place; and with very large developments, or those that set precedents by contravening established policies or means of regulation, a response among actors beyond the locality may well be initiated. These considerations draw our attention to the varying time scales over which different development processes occur, and how necessary it is for different actors to synchronize their efforts in relation to shifts in the macroeconomy, changes (real or anticipated) in public policy, and the availability of capital. Thus the development process often occurs over much longer time periods than the physical expression of change in the landscape, and following extensive negotiation. Proposals may be rejected by the local planning authority, for example, and their initiators may then feel obliged to wait for the return of propitious economic and social circumstances before pressing ahead with their original, or a revised proposal.

The discrete social demands on land and the tendency for capital to become fixed in land have produced a series of segmented land markets oriented towards different sectors of production and consumption. This segmentation arises from the interaction of the institutions of capital and the state, in particular from the regulation of markets to assist capital accumulation and to secure various social objectives. The use and development of land within any one sector is principally determined by the distinct accumulation conditions prevailing, whereas the transference of land between sectors is conditioned by the macroeconomic context and the changing patterns of demand which establish relative profitability. But various legal, politico-administrative and fiscal mechanisms have also been erected to regulate movement between markets.

The rural land-development process is thus constituted by distinct development processes for, inter alia, agriculture, forestry, industry, mining, housing, and leisure, plus a series of rules that determine, subject to relative profitability and political sanction, the transference of land between sectors (see Fig. 7.1).

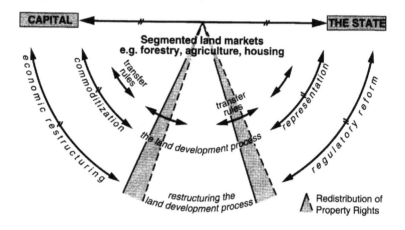

Figure 7.1 Pressures for change in rural land development.

The key rules, which play both a regulatory and legitimatory function, are those pertaining to statutory land-use planning (Ch. 5) and the fiscal regimes covering the ownership, exchange and development of landed assets (Ch. 4). Throughout the postwar period, these *transfer rules* were predicated on substantial agreement concerning the relative social priority of different demands on rural land, with agriculture accorded pre-eminence. The change and instability in rural land development that are now apparent can thus be identified in:

○ shifts in accumulation strategies and associated pressures placed on the structures of regulation within sectors; and in

○ culturally mediated societal redefinition of the countryside with its demand that the transfer rules between sectors be amended.

More particularly, as discussed in Chapter 3, the current sense of flux and uncertainty in rural land development arises from:

○ the crisis of accumulation and regulation within the predominant land-use sector of agriculture; and, additionally, from

○ government-inspired efforts to liberalize and deregulate markets (i.e. relax transfer rules), which have opened up the prospect of greater access for external capitals to rural land.

With agriculture no longer the unquestioned apex of a rural land-use hierarchy, pressures have emerged for an overhaul of the rural land-development process and its transfer rules, as well as efforts to forge new justifications to support and protect particular interests in land. Whereas dominant interests in the countryside have hitherto been successful in exercising a "power of constraint" to sustain prior commitments in the face of claims that they should be modified, the distinctiveness of the present conjuncture lies in the demand for change being articulated by those same dominant interests.

With this brief outline in mind, we review recent approaches to the land-development process, assessing their strengths and weaknesses in relation to our specific rural concern. We then attempt to refine the conceptual tools necessary for a more satisfactory approach to land development and conclude with an examination of some of the key methodological implications.

Approaching the land-development process: conceptual issues

Most of the existing literature on the land-development process adopts a much narrower perspective than the one we propose. Much of it is linked to the operation of the statutory planning system. Our concern with rural areas highlights this weakness, as the planning system is much less exhaustive in its coverage of land-use changes in rural areas than in urban ones. Agricultural development, for instance, has traditionally been exempt from development control (Ch. 5). More importantly, much of the pre-1980s literature took a narrow view of the rôle of planning in society, being more concerned with its efficiency of operation and the rationality of its processes than in the power relations that existed among those with a vested interest in the outcome (notable exceptions include Simmie 1974, Blowers 1980, Ambrose & Colenutt 1975, McAuslan 1980, Mackay & Cox 1979). More recent research has sought to redress the balance (e.g. Ball 1983, Barlow 1988, Elson 1986, Healey et al. 1988, Reade 1987), and some authors have taken as their starting point the relations between land-use planning and the multifarious functions of the state (e.g. Cloke 1989). These accounts also seek to place the development process within the context of wider political processes.

On the other hand, traditional descriptions of the land-development process (e.g. Lichfield 1956, Drewett 1973), in which the rôles and actions undertaken by actors at different stages in the process are central, have usually failed to tie the behaviour of actors into a larger analysis of the social relations of development. They acknowledge, for instance, that conflicts of interest may differ substantially according to the stage in the building cycle, local economic circumstances and alternative investment opportunities. However, their efforts are primarily directed towards developing systematic, chronological accounts of the land-development process. Key actors selected always include the planner and the developer, and frequently the "redevelopment landowner" the "exchange professionals", "financial intermediaries" and the final consumer. Less frequently included are the suppliers of infrastructure and those who do not hold a private interest in the relevant property at any stage. In the latter case, such interests are usually relegated to the rôle of "objectors" in the context of the planning process. At best these descriptions of the development process can be seen as ways of presenting ideal types; at worst they exhibit an obsession with form and sequence at the expense of process and interrelationships. This conclusion is reinforced by Barrett & Healey who, in the conclusion to their review of land policy in the UK (1985: 30), note that:

> It may be relatively easy to describe the "ingredients" or resources to be assembled for development to take place, but it is more difficult to find a way of describing the development *process* itself that encompasses the interactions of the different activities, the range of agencies involved in the process and, particularly, the complexity of their interrelationships.

A recent review of the land-development literature by Gore & Nicholson (1991) identifies four models into which previous studies maybe grouped. First, is the "sequential or descriptive" approach, which depicts the development process as a series of stages or events (Cadman & Austin-Crowe 1978). This approach imposes a rigid sequential framework on the process, thus failing to capture the diversity and flexibility that characterize development activities. However, the "pipeline" model of Barrett et al. (1978), which is perhaps the most useful of these accounts, conceives of the process as a continuous spiral, with a new pattern of development or land use emerging at the end of every cycle. This model is dynamic and flexible, allowing the relationships between its various elements to change over time and also goes some way to placing the development process in a wider context. For example, the pressures and prospects for development are linked to public-sector policies and private-sector aspirations. However, Gore & Nicholson (1991: 711) find the model inadequate in its treatment of these pressures, arguing that the "external forces remain as undifferentiated 'black boxes' on the fringes of the model, and the ways in which they influence the development process remain unclear".

The second model identified by Gore & Nicholson is the behavioural or decision-making approach, which emphasizes the rôles of different actors in the process and the importance of the decisions they "make in ensuring its smooth operation" (1991: 706). Two sub-categories of this model are described – the "individualist" and the "interactive". An archetype of the "individualist" approach is Bryant et al. (1982), which portrays the conversion of land from non-urban to urban use, identifying the main actors at each stage and categorizing them according to whether they have a direct interest in land (farmers, developers, builders, for example), or an indirect one (planners, lawyers, exchange professionals etc.). The model is seen as flawed in that it idealizes the rôles of the participants and treats the process as a closed system with little consideration of the rôle of external factors and the ways in which they might influence decisions and events at different stages. Also considered in this category is Goodchild & Munton's model of six development "routes", each passing through four decision "nodes": the allocation of land in a plan, the sale of land for development, planning permission, and the commencement of development. Gore & Nicholson view this as a flexible model, because it identifies different development trajectories, but they regard it as weak in tracing the links between the overall context set by tax and planning policies and the particularities of the development process.

The "interactive" approach, on the other hand, attempts to explore the relationships between the participants in the development process and goes some way towards linking these internal relations to those outside the process. There are, Gore & Nicholson argue (1991: 716), two central premises to such an approach: "first, that any decision or action in the development process will condition all other decisions and actions; and second, that most decisions and actions only occur after negotiation with other actors has taken place".

The work of Ambrose (1977, 1986) on the development system, for example, identifies three main "fields" – the finance industry, the state, and the construction industry – that surround the development process and condition the rôles and actions of the various participants. Here the external influences are emphasized but "at the expense of much finer detail concerning the actual processes of development" (Gore & Nicholson 1991: 719).

The same problem besets the third model considerd by Gore & Nicholson, the production-based approach. Here Boddy (1981) and Harvey (1978) are cited as examples. The core of the development process is a circuit of industrial capital, linked to financial capital, that seeks out development opportunities in order to sell them to property companies. These in turn sell the space in return for rents. The strength of this approach is that it ties together the participants in the process (and the terms of their interaction) within the overall context of commodity production. But the analysis is pitched at an abstract level, with the determining rôle being given to the needs of capital, and especially the interaction between the circuits of industrial and finance capital. This leads to the criticism that little room is to be found for human agency, choice or discretion.

If we follow Gore & Nicholson's view of this literature, an appropriate balance between the internal dynamics and the external environment of the development process is hard to achieve, although their fourth model comes closest. This is the "structures of provision" approach. Here

different types of development are characterized by different institutional, financial, and legislative frameworks, and as such the search for a generally applicable model of the development process is futile. Instead, each type of development is seen to have its own distinctive "structure of provision", whose features may be built into a separate model (1991: 706).

The prime exponent of this approach is Ball (1983, 1985, 1986a, 1986b). Once again, the institutions and agencies, both within and around the development process, are identified. Production is analyzed in terms of the social relations between agents and institutions. However, these agents and institutions are defined in relation to the "structure of provision" – the production, exchange, distribution and use associated with a specific development. Ball highlights the relations between actors and portrays the structures in which they are enmeshed as sufficiently enabling to allow discretion and choice within the development process. Moreover, the model itself is flexible because the distinctive "structures of provision" that surround a particular development process can be empirically delineated. The model simply provides an overall framework for analysis.

Healey (1991), in an approach to urban regeneration that resembles the "structures of provision" framework, argues for an institutional analysis of the development process that seeks to combine neoclassical models of the land and property markets with structural dimensions abstracted from recent political-economy analyses. In doing so, she emphasizes institutional structures and the ways in which individual actors function within them. She suggests that the connections between the actors and institutions identified as being within the development process, and those institutions and actors surrounding the process, can be followed in terms of "networks" (105). An "institutional map" could be constructed to show the nature and scope of these networks, both within and around the development sector. Healey also discusses the restructuring processes that lead to the establishment of new networks of relationships between agencies within the financial sector and between developers and local government, for example. Furthermore, she asks what is the "driving force" of these institutional relations? The answer proposed, for the current period at least, is the "changing flows of finance into and out of the property sector". Drawing upon Harvey's work (1982, 1985), Healey points to the cyclical pattern of urban development and connects it to the broader spatial restructuring of the economy, which has run in parallel with institutional change within the development sector. Although this focus on the broader patterns of economic restructuring is welcome, it fails to detail how the institutional networks are constructed and maintained.

Elsewhere, Healey, with Susan Barrett (1990), has gone some way to indicating how the collection of such detail might be undertaken. They confront the process/context dilemma by arguing (90) that any analysis of the development process requires "an explicit approach to the relation between structure, in terms of what drives the development process and produces distinctive patterns in particular periods, and agency, in terms of the way individual agents develop and pursue their strategies".

Healey & Barrett draw upon the work of Giddens (1984) to indicate how such an analysis might proceed. Here structure can be seen as the framework "within which individual agents make their choices", contains the resources to which agents may have access, and represents the rules "which they consider may govern their behaviour and the ideas upon which they draw upon in developing their strategies" (Healey & Barrett 1990: 90). In conclusion, they identify four research themes: first, the relationship between the financial system and the development process; secondly, how the strategies of particular firms are influenced by the rules and resources of their organisation; thirdly, how the state structures land and property development; and fourthly, the need to evaluate the outcomes of these processes. However, despite Healey & Barrett's plea for a "structurationist" perspective, it is still not clear how we can convincingly move beyond past attempts to straddle the process/context divide or, as it is more conventionally known, the agency/structure problematic.

This broad review of the literature indicates the need to be sensitive to issues in the process of investigating the development process. First, as we argue in Chapter 6, we must be clear how we conceptualize actors and action; secondly, we must identify how the networks of social relations between actors are constructed and maintained; and, thirdly, we must be able to explore the development process in terms that specifically identify its links to wider political, social, cultural and economic processes.

Approaching the land-development process: towards a methodology

In Chapter 6 we adopted Hindess's conception of an actor as a "locus of decision and action", where action is somehow a consequence of the decisions taken. This definition implies that actors are not only human beings; any collectivity capable of reaching a decision and acting upon it could be included. This definition immediately invalidates the position of methodological individualism. According to Elster, methodological individualism can be defined as "the doctrine that all social phenomena – their structure and their change – are in principle explicable in ways that only involve individuals – their properties, their goals, their beliefs and their actions" (Elster 1985, quoted in Bohman 1991: 148).

But Hindess, in a critique of this position, argues that although actors make decisions and act upon them, they do so under conditions that are only partly under their control and "on the basis of the techniques, ways of thinking and means of action available to them" (1988: 97). Separate actors have differential access to such tools and techniques, access that is determined by their position within a particular society. Furthermore, this undermines the conception of the unified, rational actor that underpins methodological individualism (including rational choice theory and much of neoclassical economics), for there is no reason to suppose that actors will use the same tools and techniques in all areas of their activity, or that they will be consistent in their use in all those areas. This forces us to take seriously the tools and techniques of decision making and action. As Hindess argues (1988: 109): "To dispute the assumption of rationality . . . is to raise questions concerning the techniques and forms of thought employed by or available to actors and questions of the social conditions on which they depend".

This conception of the actor forces us to situate action firmly within the social conditions in which action takes place, bringing us to the doctrine of methodological situationism. Following Goffman (1974), methodological situationism concentrates upon behaviour in its social context. Actors take decisions and act upon them, but the decisions and actions taken depend on the social context and the activities of other actors involved. In this sense, the context has a reality and a dynamic all of its own. As we emphasized in Chapter 6, we cannot simply read off this dynamic from some prior categorization of the actors; outcomes are not determined by the structural positions of the participants. Although actors may hold different amounts of resources and may have different understandings of the rules of social interaction, "rules and structural variables do not normally specify a unique course of action but are interpreted in practice against a background of situational features" (Knorr-Cetina 1988: 30). These features must be "read" by the actors, and they must draw upon the tools and techniques available to them to understand how to act in a particular context, including the activities of other actors. So the relation between the tools/techniques and their use in particular situations is doubly problematic. As Knorr-Cetina (1988: 30) puts it: "the snake bites its tail; the relation between the situation and structure (rule) is reflexive in that each is identified and elaborated in terms of the other, and the meaning of each becomes modified in the process of identification/ elaboration".

The behaviour of actors in contexts is therefore subject to some degree of indeterminacy. Yet a process implies that structured outcomes are achievable on a regular basis. Knorr-Cetina employs what she terms the representation hypothesis to account for structural regularities. Here the focus is on the representations that actors make of their interrelations. The "macro" becomes part of the situationally specific actions of the participants. It is a representation that "stands for" events within these situations. The "macro" in this formulation is not to be conceived as a layer of social reality that either "sits on top of" or lies beneath the "micro". Rather, it is firmly tied to the practices of actors in concrete situations. These practices run in parallel with one another: the outcome of some are what we call the "macro", others are forever doomed to the "micro" level. We must follow these actors, as they use their resources, implement strategies and attempt to "engineer" the "macro" or "micro" worlds in which they live.

The methodological implications of this argument are important whatever the shortcomings of the hypothesis. First, we need to examine the decision-making practices of actors and how these are then translated into action. Secondly, these practices will depend to some extent upon the decision-making practices of other participants in these situations. We must examine how actor behaviour in the development process is a consequence of "interlocking intentionalities" within specific contexts. Actors are using the tools and techniques available in order to read the situation itself, while the tools and techniques themselves are partly given by the situation. Lastly, the issue of representation entails a focus on how the actors construe their interrelations, and the tools of social research can encourage actors to make clear what their representations of their circumstances are. As Silverman (1985: 16) puts it: "once we rid ourselves of the palpably false assumption that interview statements can stand in any simple correspondence to the real world, we can begin fruitful analysis of the real forms of representation through which they are structured". Most importantly, however, we must also take account of the differential access that actors have to the resources necessary in making such representations. We need to examine the different tools and techniques that the various actors have at their disposal, their ability to use them and the way such techniques constrain the kinds of representations they can make. Representation can never capture all there is to be represented about particular circumstances. Thus "claims to represent are at the same time political strategies, potential topics and resources in the power struggles of everyday life" (Knorr-Cetina 1988: 44–5). But the issue of differential access to tools and techniques in the course of these "power struggles" is not addressed. As Mouzelis (1991: 90) points out, "she sees no difference in scale or complexity between those who have political power, and those who do not". We have discussed the issue of power at some length in Chapter 6, examining how certain categories and linkages were imposed on actors in order to restrain their freedom of manoeuvre. Sets of actors were represented by others who sought to define the interests of these others, thus restraining them from acting otherwise for as long as these representations of their interests were accepted as legitimate. Representation (in the political sense of the term) became the subject of struggle.

A socio-institutional approach, akin to that outlined by Healey (1991), allows us to focus upon the internal structures of the institutions, the means by which decisions are reached, and how these are translated into action. These decision-making procedures and the actions that follow will be partly dependent upon the situation in which this activity is taking place and therefore the range of actors in that situation. How intentionalities are interlocked should be a prime focus of the analysis, and Healey's notion of an "institutional map" could be applied to specific development case studies as we follow the actors through the process. If we follow the institutional map it will take us in and out of the development sector as we trace how categories and linkages are imposed by some actors upon others.

This analysis points to the need for a methodological approach that allows access to the dynamics of the social context and the processes of development. The most applicable method is the case study, in which a particular event or sequence of events associated with land development can be explored in depth. If we wish to follow the actors, their decisions and how they act upon them, then the case study approach is particularly appropriate as it provides an opportunity "to highlight and analyse the processes by which social actors actually manage their everyday social worlds and attempt to resolve certain problematic situations" (Long 1989: 248). More broadly, Clyde Mitchell (1983) defines the case study as the documentation of some particular phenomenon or set of events that has been assembled with the explicit end of drawing theoretical conclusions.

In this sense the case-study method would necessitate the presentation of land-development histories from an action perspective. Interviews could be undertaken among two sets of actors. One set might be directly involved in the development process; the other would be based on a survey of actors associated in some way with land development. Some developments such as a barn conversion, only involve a relatively small group of actors, such as a planning officer, farmer and agricultural advisory officer; others would be much more complex and necessitate a wider range of interviews. The final number of interviews would be dictated by the nature of development itself. The type of material generated would be qualitative, allowing as much flexibility as possible for the actors to "represent" themselves.

However, we are not concerned with case studies solely for their theoretical utility; we also wish to use them for what they tell us about the wider processes of change. As Clyde Mitchell (1983: 193) puts it: "In the analysis of a social situation some restricted and limited (bounded) set of events is analyzed so as to reveal the way in which general principles of social organization manifest themselves in some particular specified context".

The significance of individual cases can only be properly ascertained when the actors are placed in context and this context is seen, as Giddens stresses, to enable and constrain the courses of action available. More than this, the context provides the tools and techniques that allow certain decisions to be reached and certain forms of action to be regarded as legitimate. And, as Knorr-Cetina (1988) noted, in certain circumstances "the snake bites its tail", the specific alters the general; the context is modified by particular actions-in-context.

We have also stressed the rôle of representation and noted that this is accessible to the researcher. Respondents represent themselves within the interview process in similar ways to those they use to represent themselves in other social processes. These representations provide insights to the tools and techniques that actors use in the process of decision-making, and thus open internal decision-making processes to external scrutiny. In practical terms, there are obviously limits to this research methodology but a flexible research framework can allow at least some exploration of the decision-making processes.

While we may subscribe to the view that social situations have a dynamic of their own that prevents a foreclosure of outcomes, we can also attempt to control for certain features of these situations. The research context can be assessed according to criteria thought to be relevant to the behaviour in question. In the analysis of the development process, for example, such criteria might include patterns of landownership, general levels of economic activity, and the overall political configuration. Controlling for selected background variables may then allow a comparative assessment to be made of the outcomes associated with actors operating in different contexts. This approach may then allow the exploration of local differentiation in the development process. Having kept constant a set of development processes and having controlled for a set of background features, then comparisons of processes and outcomes can fruitfully be made. Case-study work can be systematic, as well as reflexive, and sensitive to the range of real-life situations. Case studies of land development thus require both contextualization and comparison. Contextualization is based on analysis of local and strategic processes that consolidate or undermine actor networks (it is worth noting that contextualizing processes derive from other actors in their networks). The processes provide the broad social environment in which action takes place. Secondly, a comparative dimension is required that allows a cross-assessment of land-development histories both between different types of land development and, at a higher level, between different localities. By choosing cases of the same type of land development in a series of localities, it is possible to develop a comparative framework for assessing local social action. While the land-development outcomes may be similar in different localities, the social context and actors' strategies involved may vary. This differentiation in social action is a potent force in contributing to the broader processes of uneven development implicit in contemporary rural change, and its analysis will also extend our knowledge of the causes of social, political and economic variation. Case studies and comparative analysis thus provide an empirical design that enables an assessment to be made of power networks at the national and local scales and the links between the two.

Towards a rural land-development process

As we mentioned earlier, our interest in the land-development process springs from a concern with examining broad patterns of change in rural areas. The approach we have adopted seeks to place the development process in its wider context and, by employing the maxim of Callon et al. (1986: 223) to follow "without fear or favour" what it is that actors do, to map the networks of actors engaged in rural land development. But in what sense can land development be regarded as rural?

There are two responses to this question. First, there are certain development processes that can only be accommodated in rural areas because they are extensive users of land. These include agriculture, golf courses, and mineral extraction. Secondly, there are other development processes that may take distinctive rural forms, though not exclusive to rural areas, such as housing.

A provisional listing of the wide range of processes and their outcomes that are frequently found in rural areas is presented in Figure 7.2. This shows how the agricultural land-development process, for example, can incorporate a quite varied group of actors. For practical reasons, the processes are defined on sectoral grounds because each sector has its own policy context, market conditions, economic structure and set of statutory rules, amounting to something akin to Ball's (1983) structure of provision. However, we should expect with the restructuring of rural areas, both in terms of the competitive position of alternative uses of rural land and of the 'transformation' rules mediating between them, that it is the interlinkages between these sectors that are of growing importance.

This point is illustrated in Figure 7.3 in relation to just two sectors: agriculture and housing. The potential range of strategies of the different actors and the degree of their involvement is considerable, although the terms of their participation will depend on the activities of other actors and their relationships to sets of resources: property rights, local economic conditions, finance, and knowledge (of political/planning procedures in particular). But it must be stressed once again that we cannot define in advance the rôles and interests of these actors. We must track through the land-development process the strategies adopted by the various actors as they attempt to define development problems and solutions to those problems, as they try to enrol other actors to their cause, fix the identities and interests of those others and impose their own perceived interests. This process will take place within a dynamic social context that enables certain courses of action and constrains others. These preconditions are partly the result of past development pro-

Figure 7.2 A sample of rural land-development processes.

Sector	Type of outcome
Agriculture	*Intensification:* e.g. drainage scheme, grassland to arable, new farm buildings. *Extensification:* e.g. fallow. *"Manufacture":* e.g. vegetable packing; cheese manufacture based on farm-produced goods. *"Retailing":* e.g. pick-your-own; farm shop.
Housing	*"Family needs" provision:* extension, or subdivision of existing property. Reproduction of the non-family workforce: tied accommodation, housing for hired labour. *"Small-scale" production:* (i.e. infill, up to 0.5ha) e.g. barn conversion; houses (new structures) [both for sale]. *"Large-scale" production:* (>0.5ha) housing estates (private); housing associations + housing estates ("welfare", including trusts, sheltered housing, other forms of public provision).
Accommodation	Guest houses (conversion) Mobile homes Camp sites
Recreation/leisure	Golf courses, country parks, heritage facilities (museum, stately home), country clubs, time-share apartments, holiday flat, picnic sites, equestrian facilities.
Mineral extraction	Gravel, open-cast coal, chalk/limestone
Woodland	*Afforestation:* e.g. amenity woodland, farm forestry, commercial sylviculture.
Commercial	*Industry:* factory buildings, workshops, offices, warehousing, retailing (out-of-town shopping centres).
Infrastructure	*Public utilities:* reservoirs, substations, pipelines and power lines, sewage farms, refuse dumps.

cesses. Present and future outcomes (or non-outcomes) will result from the complex interaction between 'action' (understood as social relationships becoming fixed within particular power configurations) and 'context' (understood as a shifting distribution of economic, political and cultural resources).

As we have stressed, we are concerned to "follow the actors" as they make decisions and act upon them, as they attempt to

Figure 7.3 Actors engaged in various rural land-development processes.

Outcomes	Agriculture			Housing			
	Drainage scheme	Extensifi- cation	Vegetable packing	House extension	Barn conversion	Sheltered housing	Private housing estate
Farmer (o/occ)	X	X	X	X	X	X	X
MAFF		X					
ADAS	X		X				
NFU				X			
Accountant /solicitor	X	X	X	X	X	X	
Bank manager	X	X	X	X	X		
Building society				X	X	X	X
Financial institution						X	X
Builder/ contractor	X		X	X	X		
Developer						X	X
Estate agent				X	X		
Local planner		(X)		X	X	X	X
Local housing department						X	X
DOE							X
Parish council		X			X	X	X
Rural commission council						X	X
House buyer				X	X	X	
Amenity society	X					X	X
Rural development commission			X				
Public utility		X		X	X	X	
Local residents		X	X	X	X	X	

influence the operation of the process at its different stages or its final outcome. Furthermore, it would be useful to have some means of assessing both the distributional and reproductive consequences of the processes. The detailing of financial returns to the different parties would be one means of assessing who gains and who loses. However, this has rarely been attempted because of the difficulties of obtaining reliable information. An alternative approach would be to focus on the redistribution of property rights between actors at the micro level as the process of development unfolds. It provides one means of measuring the social consequences and changing power relations. The distribution of property rights can indicate the changing relative strengths of the different actors: their ability to impose their representations upon others and determine the scope for alternative courses of action by their competitors. However, the detailed examination of the redistribution of property rights has to be accompanied by the collection of other evidence on the objectives, activities and returns that each actor expects from the development process. Formal property rights cannot in themselves be more than an initial indicator of the control actors can exert over the process, or even the objectives of actors, but the decision as to whether to retain the freehold interest or to sell it may also provide a crucial insight into the expectations and goals of particular actors.

Conclusion

We have sought here to identify some of the key methodological priorities and prerequisites of the study of the rural land-development process, given our earlier conceptualizations. Our survey of the existing literature on land development reveals that although many studies display a combination of elements integral to the process/context and agency/structure debates, none has resolved this conceptual difficulty. We therefore returned to aspects of the discussion contained in Chapter 6 which requires that our attention be focused upon actors in specific contexts, and the types of strategies they regularly pursue, in order to understand land-development outcomes in varying social situations. Actors' strategies, and the power configurations linked to them, are inherent within the land-development process, and it is only through an understanding of these that the physical outcomes attributable to development can be accounted for.

CHAPTER 8

Constructing
the countryside

Introduction

It is now time to re-assess our arguments in the light of the questions posed in Chapter 1. Those questions concerned both macro and micro issues to do with rural change and its regulation and, in particular, the ways in which international processes of economic and social restructuring are being expressed within national and local contexts. They were, first, how should advances in mainstream social science theory be applied and adapted to the rural arena? Secondly, to what extent do the regulatory and accumulation "crises" of the late 20th century suggest a significant break with past experience, and with what consequences for rural areas? Finally, how should locally based social action be incorporated into our understanding of uneven development?

The contemporary shift in the social, political and economic conditions of rural areas is of sufficient magnitude to demand a revised conceptualization of rural change. A seductive avenue for researchers has been an analysis of the processes of internationalization (both political and economic) and their potential spatial consequences. But top-down empirical demonstrations of the consequences of international tendencies for national, regional and local structures, which begin with descriptions of global tendencies and attempt to predict local responses, remain problematic. More compelling, as Chapter 6 contends, is the reverse approach: seeking evidence of local action and local systems of relationships in the formation of rural localities in a more internationalized world. This perspective challenges what we regard as a former, unreflexive application of structuralist concepts to rural change in which the distinctive rôle of locality and rurality in the economic restructuring and urban experience of society in the late 20th century was all too easily dismissed.

In seeking to address these issues, our thesis has moved from the abstract to the concrete and from the macro to the micro. It has pursued this course as a way of analyzing the political, economic and social processes that shape rural localities. After reviewing and re-assessing some of the key conceptual tools necessary to the study of rural change (Ch. 2), we have attempted to apply them to a particular case – the evolution of rural space within the UK since the mid-19th century – and to do so by placing them within a broad social regulationist perspective. This approach has allowed us to analyze the relationships between production and consumption in rural areas and to focus upon how these spheres of activity are regulated. In particular, we regard the regulation of property rights and the ways in which they are commoditized and represented as key mechanisms in influencing both the rate and direction of change. We have attempted to demonstrate the insights this perspective can bring by making an historical analysis of the changing regulatory strategies adopted by the national and local organs of the British state towards its rural areas, paying particular attention to the last 50 years. From this review, we feel able to conclude that the present period does indeed represent a significant break with what has gone before. We suggest that this arises in part from the gradual decline of an international, Atlanticist food order, as illustrated by the faltering progress being made towards reform of the European Community's Common Agricultural Policy and of international trading conditions under the Uruguay Round of GATT negotiations. In the UK, that food order sustained, and was sustained by, the primacy given to food production in both domestic agricultural and planning policy. Specifically, the protection of farm land and the containment of urban growth dovetailed with the aims of stability in commodity markets and prices, and the quest for efficiency in agricultural production.

It is important to recognize that this mode of regulation provided more than a consistent institutional and policy framework that defined national rural space. It also set particular social conditions – including the release of agricultural labour and the necessity for farmers to adopt increasingly sophisticated technologies and a real increase in the price of property – with consequences that included the progressive middle-class takeover of rural living space and the growing dependence of farmers on agribusiness. Value was thus conferred on the production of agricultural commodities and, by default, on those people who controlled the means of their production, as well as those fortunate enough to gain access to rural living space but who were often not dependent on that space for their livelihood. It was a mode of regulation based on a highly centralized system of state inducement, and it was maintained by a political and social consensus that lasted for over 40 years until the 1980s. It is now in variable degrees of retreat in almost all advanced capitalist economies. At one level, redundant agricultural workers were progressively absorbed into an expanding urban economy, creating social gaps in rural life frequently filled by a quite different and acquisitive "middle class" that set about altering local social and political agendas; at another level, powerful agricultural interests ensured the continued payment of support monies to farmers in ways that sustained profits in the food chain while presiding over the growing economic vulnerability of the traditional family farm.

By the mid-1970s, however, some commentators were proclaiming the breakdown of these postwar conditions – not so much in the agricultural sphere, protected as it was by extensive national and international state management – but much more widely in terms of the ending of a postwar economic boom that had been manipulated and sustained by the trading and financial power of the USA. Oil "shocks", unstable currencies and rising interest rates all fed upon each other to create a loss of confidence, a retreat from free trade and, ultimately, a loss of faith in a Keynesian approach to macroeconomic management. These concerns, combined with those over the future of the global and local environment, precipitated a reassessment of political and economic priorities and the means of achieving them. Together, they led to a general drift towards neoconservative thinking and an attempt to develop a new social mode of regulation that triggered, among many other consequences, a process of change in rural space and society. For example, it fuelled the gathering critique of agriculture's primacy in rural land use, the single-mindedness and protected status of agricultural policy-making, and the existing forms of agricultural corporatism. It began to unhook some of the links in the food chain that had, hitherto and to varying degrees, attached particular rural localities to a rapidly globalizing economic nexus. And whether the debate focused on farm surpluses, or diversification and alternative land-use, or the maintenance of traditional tenancy agreements that sustained "insider" farm families, change was always being directed towards the need to unleash new exploitable opportunities and to attract a wider range of economic interests into the rural arena.

In academic circles, and to a certain extent in those involved with making policy, the axis of debate shifted from a concern with the competitive efficiency of agriculture and other land uses versus the extent and continuance of rural deprivation, to the potential for agricultural and rural diversification versus different forms of environmental protection and regulation. These fissures in debate became especially evident by the mid to late 1980s in the different positions adopted by agricultural economists and rural sociologists (Lowe & Bodiguel 1990). In the policy debate they became visible in the different national postures being struck over the future rôle and management of rural areas within the European Community, and especially that between the UK and its continental partners (see, for example, the discussion in the House of Lords Report (1990) on the EC Green Paper *The Future of rural society*), and within the current Uruguay Round of GATT negotiations over the nature and level of protection that should be afforded agriculture in western Europe, Japan and the USA.

With the broadening of objectives towards food and fibre production, agricultural economists have been forced in their research, like farmers in their practice, gradually to relinquish their single-minded pursuit of the means of raising agricultural productivity and efficiency, and the accompanying value-laden assumptions under which the resources of capital and land are prioritized over those of labour and the interests of the consumer. As a neoconservative mode of social and economic regulation has gained ascendancy, rural sociologists have also had to refine some of their political-economy assumptions. As the structural dominance of a productivist agriculture has receded, the remoulding of local conditions by local actors and agencies has become more diverse and subject to different forms of conflict and negotiation. This process is inevitably protracted, conflictual and spatially uneven, but of particular salience is the increased emphasis placed on the derivation and allocation of property rights, both generally in society and more specifically in rural areas. They are now a major linking mechanism between the re-regulation of markets and state policy, and reveal the varied ways in which individuals, social groups and institutions secure, defend and challenge access to land and natural resources and continuously redefine them in terms of their monetary and non-monetary values. Given the spatial fixity of property rights, it follows that the *local* construction of these values is an important arena of contestation, leading to the uneven development of various markets for rural goods, such as minerals, housing, recreational activities and certain types of environment.

Neoconservative policies have ostensibly been oriented to freeing the means of production from the shackles of corporatism, a task made the more urgent by the rising cost of supporting the welfare state and other economic sectors in public hands. An additional, more purposeful, goal has been to extend private-property rights (whether in terms of owner-occupation of housing, private rights to education and health, privatized systems of transport, greater individual share ownership and more varied rights to personal consumption) as a means of stimulating the economy by providing the kinds of goods and services the burgeoning "middle class" wishes to acquire.

A highly individualized notion of land rights in a fully commoditized world clearly has limits. The historical record indicates that long after the introduction of capitalism, the social construction of the countryside has accommodated the maintenance of selective and often very exclusive rights *within* either a paternalist or corporatist ideology. The current promotion of a more overtly individualistic and populist pattern of entrepreneurship and consumption requires its own local and national forms of legitimation. To realize this, new forms of social regulation have to be *constructed* and the manner in which this takes place will vary over space and between the different development processes to which land is subject. Both the rights to access and to develop land, and the rights to trade them, are beginning to be regulated in different ways according to the regional socio-economic and political context and to the different power bases of the key actors and agencies involved.

As we discuss in Chapter 5, in the UK these underlying social and political processes have implied reforms not just for agricultural policy but also for postwar land-use planning policy, which had subordinated a more general concern for rural problems to the specific needs of agriculture. The emergence of neoconservative views on the regulation of production has combined with a growing local environmental concern to challenge the ideological basis of a set of centralized planning and agricultural policies. Whereas the social value of pursuing individual needs, including rights to land, housing and the environment, had been strongly encouraged in the early 1980s, by the end of the decade their cumulative consequences were all too evident, and were even the subject of increasingly critical scrutiny among at least some of those who had initiated the process. Specifically, once these consuming groups had acquired their new rights to exploit rural resources they needed to ensure their retention through a defence of their new social and political positional status. Thus land, and the rights conferred upon it, once again emerge as peculiarly distinctive in rural development. As with agricultural production, these consumption rights are fixed in place and are finite in extent. The need for these new land rights to become *remonopolized* under a post-productivist phase fuelled new social conflicts. In the context of a less centralized land-use planning system, the importance of the local arena of social action had once again been raised.

While remaining under a regime that encouraged ever greater quantities of food, farmers were required to adopt new technologies in order to survive, wrestling with the need to remove, as far as possible, the obstacles presented by land to production. This necessity encouraged management practices opposed by a non-agricultural population – residents or visitors – which, once it had obtained greater rights of ownership, occupation and use to rural land, wished both to express these new found rights through social and political action, and to secure the distinctiveness and authenticity of its land-dependent possessions. Far from wanting to diminish the fixity and finite nature of their land rights, they wished to flaunt and exploit them. Moreover, they increasingly expected the local land-use planning system to assist them in doing so. With the rural working class politically marginalized, the emergent "middle class" and an increasingly debt-laden farming sector have been left to contest the opportunities that locally fixed land rights engender. Partly because of these pressures, and partly because of the inadequacy of existing, sectorally oriented powers associated with agriculture, housing and so forth, the behaviour of local planners and the social forces guiding the local regulation of land-use have begun to play a much more significant rôle in determining the representation of interests. It is at this sharp end of regulation that the attempts by developers to convert existing use, occupier and owner rights, and thereby to begin the process of redefining agricultural, housing and industrial priorities, are especially evident. Central government, through the issue of circulars and guidance notes, may set parameters, but it is local action that often determines how these are interpreted and implemented. It follows that, in these particular circumstances, the deregulatory tendencies of the central state have sown the seeds for a more uneven and more localized pattern of *re-regulation* of rural living. For the specific case study of the UK these conclusions can be examined further by reference to recent and evolving changes in the regulatory structure.

Reconstituting corporatism: towards a new regulatory order in the British countryside?

In the UK, the corporatist relationship between the farming lobby and the Ministry of Agriculture, Fisheries and Food traditionally cast its thrall over a series of subsidiary agencies, including local planning authorities, the Nature Conservancy Council, National Park Authorities and the Forestry Commission, ensuring that in their land-use planning functions the needs of agriculture were safeguarded. This institutional hierarchy embodied a set of priorities for rural land in which food production was pre-eminent, and acceptable subsidiary uses (including nature conservation, recreation and forestry) were accommodated at the margins. Agricultural corporatism was thus the keystone not only of agricultural policy but also of postwar rural planning. The 1980s, though, saw a waning of agricultural corporatism as the productivist programme of agricultural expansion was beset by many problems. In addressing these problems, ministers' deregulatory instincts, and a wish to wean farmers off their dependence on price supports, coincided with efforts to liberalise the planning system and to open up the countryside to new forms of investment. The consequence has been to break the hold of agricultural corporatism over rural planning, leading to considerable uncertainty in land-use management and a policy and political vacuum in the planning of rural areas. Other interests – minerals, housebuilding, waste disposal, leisure, forestry and environmental – have sought to exploit this vacuum directly, by gaining access or control over rural land, and indirectly through efforts to establish new, or break into existing, corporatist structures, such as the Annual Review of Agriculture, Regional Aggregates Working Parties, Housing Land Availability Studies and Regional Forestry Advisory Committees. At the same time, many local interest groups have tried to take advantage of the "opening-up" of the countryside to establish localized regulation through the planning system.

Individually and locally, of course, farmers and landowners, by virtue of their control over rural land, remain as significant private interest regulators of changing land use and development; but the farming "lobby" is presented with a major dilemma in its uncertain moves away from an agricultural preservationist stance towards one more in favour of rural development. These moves run counter to ingrained ideologies of productivism and stewardship and generate considerable internal strains for the farming community. Development on one farm can create pressures and disruption for neighbouring farmers. Diversification may be an economic panacea for certain farm households, but can provoke resentment among those others striving to farm as "normal". Moreover, the renewed emphasis on land as a capital asset – including production control payments to farmers such as milk quotas and set-aside – reopens old tensions between productive and landed capital. Although many farmers may be ruing the fixity and cost of productive capital assets, (most notably those who have bought on the open market since 1980 when the real price of farm land has been in almost continuous decline and real interest rates have been at historically high levels), they now have selective opportunities to convert that previously fixed productive capital to other uses. Attempts to exploit these opportunities frequently arouse opposition from other actors inclined to treat with scepticism farmers' claims to be the guardians of the countryside.

It is important to recognize, though, that as one traditional corporatist determinant of rural planning retreats, others are emerging into greater prominence. In the building, "digging" and "dumping" businesses there has been considerable corporate concentration, with a small number of large firms coming to hold extensive capital and land resources and to dominate the sectors of construction, minerals and waste. Major firms of house-builders have assumed an increasingly dominant position in the housing land development process, ranging from land acquisition and assembly to construction and sale, as well as encompassing the entire development of new settlements. Through the scale of their operations, including their ownership of, or options over, extensive areas of developable land, and their ability to provide infrastructure and capital, the cooperation of the major building firms has become vital to the realization of housing targets in many local plans. Indeed, they have come to enjoy a corporatist relationship with the planning system, mediated by the Department of the Environment, its regional offices and the joint Land Availability Studies. Where they have wanted it, these arrangements have provided corporate housebuilders with a key consultative rôle in planning the supply of housing land; there is plenty of evidence to suggest that the housebuilding sector as a whole wants the retention of clear and firm planning strategies as these reduce the level of uncertainty in their own operations.

Thus, although the government has consistently urged local planning authorities to cooperate with private housing developers, for their part, the housebuilders, very much like the farmers, have been ambivalent about moves to deregulate the planning system. They have certainly not wished to see the wholesale relaxation of planning control since this would destroy the value of their land banks. Rather, they have pressed for a more "flexible" and responsive system of local negotiation and land release. This approach has further reinforced the position of the large, well resourced and skilful developer, and has highlighted the differentiation of local forms of social and political regulation. Conscious of the positional good element in rural housing consumption, housebuilders have been concerned to ensure the controlled release of land in appropriate places. Their decisions on the phasing of the development of this land and its stratified allocation – whether, say, as "executive" or "starter" homes – play an important part in the social reconstitution of the countryside, as do their attitudes towards housing associations and the supply of affordable housing. The corporate housebuilders, whatever their free-market rhetoric, want sufficient but not excessive amounts of land released for development, and they want the bulk of it for themselves. Increased competition would cut into their profit margins. They seek a compliant planning system, not a deregulated one. Exclusivity in terms of both production and consumption thus eclipses issues of equity in the rural housing land-development process, a balance of priorities aggravated by the extremely limited amount of new low-cost housing that is being built and the disproportionate loss of housing stock from the public sector in rural areas as a consequence of "right to buy" legislation (Shucksmith 1990).

The rôle of the housing land-development process in the social reconstruction of the countryside shapes, and in turn is shaped by, local environmental concerns. Such concerns have been most actively articulated by the successive waves of ex-urban, professional and managerial middle classes, attracted to and seeking to protect the residential, recreational or amenity value of particular places. No longer is their impact restricted to the Green Belts and outer metropolitan areas, as once described by

Ray Pahl (1965) in his classic study. It is also now a central feature of the "deep" countryside. Commuter hinterlands have vastly increased in size, pushed outwards by the regionalization of urban housing markets under the pressure of high house prices, and facilitated by the extension of the motorway network, increasing car ownership, new work practices associated with developments in telecommunications and information technology, and the creation of many small businesses, as major companies restructure their operations. Retirees and second homers have often been the harbingers of this rural embourgeoisement, seeking out cheap property to gentrify, or following the route maps of the housing developers into more distant areas. The greatest pressures, though, have been experienced in country towns and villages accessible to urban labour markets and set in picturesque countryside. Earlier waves of middle-class newcomers have now firmly established themselves in rural society, recreating it by participating in, and in some cases dominating, various local social and political institutions, including the planning system.

Their actions in protecting the environment and the exclusivity of rural areas are in a reciprocal, but ultimately contradictory, relationship with those of the housebuilder. Like the latter, they seek market closure through the operation of the planning system, especially through the pursuit of such restrictive planning designations as village envelopes, Conservation Areas, Green Belts and Areas of Outstanding Natural Beauty in development planning and development control. This strenuous defence of rural space reinforces its value as a positional good that in turn fuels the market for rural housing while it ever more constricts that market. This contradictory dependence of the housebuilder on social mechanisms of market closure leaves them marketing "exclusive" housing while railing against "nimbyism" in the planning process. The main beneficiaries are those rural landowners fortunate enough to be allocated scarce development rights.

To say that there are heightened tensions between the three major interests in rural land – farmers, developers and middle-class residents – is indeed a truism. Each occupies a powerful position in the land development process: farmers and landowners through their ownership of rural land; the middle classes through their prominence in local government and politics and their property rights (owned and claimed); and the large developers through their oligopolistic control over house building and other major development processes, as well as their government-sponsored corporatist links with the planning system. What characterizes the most recent period, however, is the reconstitution and recomposition of these interests such that they rarely align themselves consistently over time or do so collectively at the local level.

Such changes are especially well illustrated by a recent study of the provision of industrial units on farms in Buckinghamshire during the late 1980s (Marsden et al. 1991). In the middle of the decade the discourse on farm diversification and the rural economy emanating from central government led the local planning authority to look favourably on this form of development. This was encouraging to those farmers and landowners who were experiencing a decline in agricultural incomes and rents, and were in a position to take advantage of a buoyant local labour market. But no sooner had farmers begun to submit planning applications to redevelop their properties to meet the needs of small light industrial businesses than a shift in central and local government thinking occurred in favour of rural conservation, made under pressure from middle-class environmentalists, and this led to new obstacles being placed in their path. With a renewed emphasis upon the siting of such units and their aesthetic quality, the development of the new units came to be much more closely regulated, and in many cases refused. This change in policy represented, at a broader level, the ascendancy of certain aesthetic representations of the countryside over economic ones, attributable in part to the changing mix of social groups laying claim to rural space both nationally and locally.

In each locality, different actors and agencies will hold the initial property rights and knowledge bases necessary for the ability to trade, while the local regulatory structure (local planning policies and plans, public utilities, conservation agencies, etc.) will bestow or moderate those rights to varying degrees. For mineral developments, barn and industrial conversions and for golf courses, for example, local planning decisions will reallocate the rights to trade between actors and between places, but in the increasingly negotiative context in which development takes place, trading in knowledge, and the holding of specialized knowledge skills, is of growing importance. The loss of real or perceived control, when holders of property rights begin trading in unfamiliar markets, is exacerbated in recessionary conditions, which drive down capital values, raise interest rates and reduce incomes. Moreover, the growing commitment of the new middle incomes. Moreover, the growing commitment of the new middle class to environmental protection, often with residential or visiting "rights" rather than productive interests in property, has placed a brake on the attempts of farmers and developers to realize new economic opportunities, highlighting the fact that all rights to trade have to be socially sustained.

These revised social and political representations can also be seen to reflect a series of new tensions in the evolution of party politics, particularly but not exclusively, at the local level. As Lash & Urry (1987) have argued, the traditional constituencies of an urban-centred Labour Party no longer automatically align with working-class or middle-class thinking, and so too, somewhat later, we may be witnessing a fracturing of the traditional rural Conservative vote. As with the archetypical trade union Labour voter in manufacturing industry, some of those groups who have most consistently supported the Conservatives are now losing economic and political power. English farmers may be inclined to remain Conservative through thick and thin, but their electoral and interest-group power is on the wane. Moreover, again like many of the traditional trade unions, farmers have become less united and more prone to intra-group conflict and bouts of political exasperation. The vote of no confidence in the Minister of Agriculture by the 1987 annual meeting of the National Farmers' Union was unprecedented, whereas down on the farm, farmers have increasing difficulty justifying their activities to the parish and district councils that they no longer necessarily dominate.

The new rural middle classes have exhibited, too, a curious relationship with the Conservative Party. They are supportive of its national policies and governments, as reflected in the results of the past four general elections (1979, 1983, 1987, 1992), but less reliable in their commitment to it in local and European elections. Although infrequently prepared to support the Labour Party, rural voters often elect candidates from other parties who may be conservative in outlook but not convinced of the merits of New Right policies for their areas. Their willingness to do so has been most apparent when the negative consequences of central government action for local services have been to the fore. The disaffection of the ex-urban middle class was most clearly revealed in the 1989 elections for the European Parliament. Evidently, many of them had psychologically detached themselves from their urban and suburban roots, and like their predecessors analyzed by Pahl, were "going local". In that election, the Green Party, with its slogan of "think globally, act locally", polled over 2 million voters, or 15 per cent of the votes cast, and came in second or third place throughout most of non-metropolitan England. Although it benefited considerably from the collapse of the alliance of the two centre parties, it drew its greatest support from former Conservative voters (29 per cent of the total, according to post-election opinion surveys). Throughout southern and eastern England this bit deep into Tory majorities. The increasingly amenity-minded middle classes were revealed as being both highly fractured and unevenly reconstructing their rurality around different collections of positional goods. As some backbench Conservative MPs were to discover, they were, and indeed still are, quite capable of defining the political and social *boundaries* of neoconservatism, rather than necessarily upholding its central motive forces. If "suburban man" typified the genesis and direction of Thatcherism in the early 1980s (Jacques & Hall 1989), reconstituted "rural man" had come to represent its limits by the end of the decade.

The evolution of these complex forms and fissures among social groups in rural localities represents more than a tendency towards the refracturing of social classes based on evolutionary notions of post-Fordism, consumer orientation or disorganized capitalism. What they do appear to represent, however, is the singular *lack* of a coherent form of social regulation to match the new forms and direction of economic activity impinging on rural society (Peck & Tickell 1992). Rural society in the UK does not represent a coherent social system that allows for the reproduction of a comprehensive rural economy. Instead, we have to focus upon both the variegated spatial impact of social and economic restructuring and its coincidence with different local systems of regulation and social relationships bound up with the land-development process.

Part of the reason for this lack of coherence in the social regulation of the UK's countryside stems, as we have seen, from the economic and political processes "running through", as well as being actively moulded by, rural society. It is also intertwined with the declining class-based orientation of modern society, the increasing significance of national and local knowledge systems, and the construction of identities and action in local and regional contexts (Ch. 6). These contexts are created by sets of strategic and local interests and their representations, and are leading not to a new and exclusive mode of social regulation of rural space but to the construction of differing local and regional forms.

In rural Britain, the growing importance of local conditions stems not only from capital's search for greater mobility and the increasing adoption of more flexible systems of production – processes evident in all advanced economies – but also from the social and political consequences of the Thatcherite experiment itself. Thus, although the experiment sought to establish a new mode of regulation by, for example, altering consumer expectations, opening up individual rights, and decollectivizing workers' rights, in the rural domain it has failed to develop any coherence because of the social fracturing of the increasingly dominant middle class. Instead, locally diverse social modes of regulation are beginning to typify rural society, variably resisting the underlying logic of current forms of accumulation. These circumstances differ significantly from those that we outline under the heading of the Atlanticist and productivist regime in Chapter 3. At that time, the prescriptive and centralized policies associated with a Fordist stage in the development of the food system were sustained in the UK by complementary planning policies, and together they provided conformity of national purpose, legitimation and stability. None of these ascriptions are so evident today.

Towards the differentiating countryside

It will be evident from the preceding analysis that we consider the current processes of economic, political and social change to be engendering a period of flux and differentiation in rural areas. However, while the pace of change seems to be increasing, it is not taking place in a vacuum. History delivers structures and modes of behaviour. The forces for change that we have identified coalesce with, and arise out of, stable forms of social life. Change does not chaotically ravage the rural landscape. It occurs along structured pathways and provides us with the means to assess how the past acts as a context for the present.

The decline of the postwar certainties, most notably agricultural productivism and its corporatist structures, has opened the way for the emergence of a more differentiated countryside, one whose trajectory is no longer determined to the same degree by the fortunes of a single industry but by a much more complex assemblage of economic, social and political elements. These elements may be present at the local, regional, national or international scales, but will give individual rural areas quite different complexions. While there may in principle be an infinite number of these complexions, we feel it makes sense to try to simplify the position by identifying certain features that are likely to be crucial in structuring the course of future countryside change.

In Chapter 6 we drew upon the work of Massey (1984) to suggest how, within localities, new rounds of investment articulate with the old. The shift to a more differentiated countryside will, in many areas, result in increased competition for rural resources from a variety of economic actors, while in others it may mean an increased reliance upon traditional rural industries, such as agriculture. In spite of the overall decline in importance of agriculture, there can be no doubt that some localities will continue to be even more closely linked to the food industry or to biomass production, where, in all probability, new biotechnologies will be deployed in order to retain the industry's local international competitiveness. Other localities will retain a more traditional farming character where the social composition of the population may remain relatively unaltered, while in others again a shift in the economic structure may be accompanied by increased social mobility, especially where more general diversification in the rural economy is linked to the in-migration of a commuting population. Likewise, these changes can also be expected to be reflected in the types of politics practised, creating a contrast with those localities where a lack of economic and social change may allow the continuance of established political forms, such as corporatism and clientelism.

We would identify four main sets of parameters crucial to the developmental trajectories of rural localities. These consist of:

(i) *economic parameters*, most notably the structure of the local economy – its buoyancy and diversity (rate of growth, level of unemployment etc.), and the rôle of the state in the local economy (level of dependence upon state agencies and state financial support etc.);

(ii) *social parameters*, including demographic structure, rate of population change, influence of the "middle" class, level of commuting and proportion of retirees;

(iii) *political parameters*, including ideals of representation (who is a legitimate representative); forms of participation (e.g. level of interest-group activity); type of politics (whether, for example, organized around production interests or the protection of positional goods); and

(iv) *cultural parameters*, including dominant attitudes towards property rights (e.g. land as heritage (stewardship), as productive asset, as fictitious asset) and a sense of locality/community ("belonging").

The relations between these parameters are by no means determined by the economic. Political regulation can significantly influence the economic activities that take place in particular locations, as through the decisions of the land-use planning system, and political regulation is firmly tied to the social constituency. To understand the contours of rural change we must analyze the interactions between them. To give some indication of how we think these might come together in rural areas at the present time we outline below a set of "ideal types" indicating possible combinations of economic, social and political forms. The four identified do not constitute an attempt to cover all kinds of rural area; instead they seek to encapsulate major tendencies. In this sense, they do not refer to specific places; rather, variations upon them overlap and merge into one another in rural space. To seek to map them as discrete categories would be to misrepresent the purpose of identifying ideal types. The purpose is to provide an initial organizing framework within which the relations between key social, economic and political processes may be examined in particular places. Finally, it should be said that these ideal types are distilled from British experience but, insofar as they encapsulate general tendencies of rural restructuring, they may resonate with the experience of rural areas elsewhere.

The preserved countryside

Throughout much of the English lowlands, as well as in attractive and accessible upland areas, anti-development and preservationist attitudes dominate much local decision-making and political organization. This is expressed mainly by fractions of middle-class residents oriented towards the protection of amenity. Increasingly, these conditions impinge, primarily through the planning system, upon farmers and landowners seeking either to diversify their businesses or to intensify the use of their land. For many farmers, when they are located in vibrant labour market areas, farm pluri-activity involves both the progressive intensification of farm operations *and* the diversification of enterprises (Marsden & Williams 1992). In addition, demand from middle-class fractions provides the basis for the provision of new recreational, leisure or retailing enterprises. Through the local political system, though, strong opposition is likely to be

aroused by agricultural intensification that damages the environment, major developments or infrastructure projects with a
significant impact on the landscape, and any nuisance-generating
developments (e.g. "war games" or land fill). We might label
such areas as the *preserved countryside* where the reconstitution of
rurality is highly contested and often controlled by articulate
consumption interests that use the local political system to
protect their positional goods. In those areas the combinations of
property rights become highly complex and fluid, partly because
they form the basis for negotiation within a local, professional-
ized regulatory system.

The contested countryside

In areas of the countryside beyond the major commuter catchments and of no special environmental quality, farmers and
landowners may be more than simply the agents of a productivist agriculture. They may still play a leading political and
economic rôle, even if this is no longer a dominant or unchal-
lenged one. Under these social conditions, the scope for diversification may be less, but so also will be the opposition to agricultural intensification. Development projects unwelcome in the
"preserved countryside" are likely to cause much less contention,
and coalitions between farming and development interests –
often dominated by small businesses rather than large corporate
concerns – may well dominate the rural land-development
process. We might label such areas as the *contested countryside*, in
which a continuing sense of localness, rather than professional-
ism, determines the legitimacy of local representation. Decision-
making reflects local economic priorities but is increasingly being
challenged by incomers who reflect the positions of the middle
class articulated so effectively within the "preserved countryside".

The paternalistic countryside

In areas of the countryside where large private estates and big
farms still dominate the land-tenure pattern the rural land-
development process is still shaped significantly by the interests
and outlook of existing landowners. With agricultural rents and
incomes falling they may well be looking for ways of diversifying
their activities and perhaps making development gains. Many of
the large estates faced with falling rents and less enthusiasm for
"taking land in hand" have begun to sell off assets, particularly
those associated with farm buildings and estate housing.

Nevertheless, unlike the struggling owner-occupier, their consolidated ownership of land, and perhaps their continued political influence, allow them to be much less dependent upon external developers. They are still able to take a broader and longer-term view of the management of their property. Today, their management practices may reflect an uneasy relationship between financial imperative and social conscience leading to the retention, at least in certain pockets, of a paternalistic social system. This is now not so much based around a thriving "occupational community" (Newby 1977), as one where landlordism has recreated its compliant social conditions around established non-agricultural residents. We might term such areas the *paternalistic countryside*, and expect to find within it the reluctant adoption of environmental management policies (e.g. in Environmentally Sensitive Areas), where similar practices cannot be self-funded, a commitment among large landowners to affordable housing, and a local political system still influenced by representations based on landed status and long association with the local area.

The clientelist countryside

In certain remoter rural areas, agricultural corporatism in its productivist form retains a considerable hold. These areas are yet to experience the social transformation arising from the forces of counter-urbanization, and the social welfarist model of agricultural support and rural development is still prevalent (Smith 1992, Murdoch 1992). Under these circumstances, processes of rural development remain dominated by clientelist relations between the interests of farming, landowning and local capital on the one hand, and state agencies involved in promoting primary production or rural development on the other. Systems of direct agricultural support, before the current shift in European Community farm policy (e.g. Less Favoured Areas, Sheep Meat Subsidies, Hill Livestock Compensatory Allowances) have continued to encourage the restructuring and modernization of traditional farming, even though local sympathies towards income support rather than price support would be frequently expressed. Although pluri-activity is a well established business strategy on many farms, the scope for additional, locally generated economic diversification is very limited, and what there is usually depends on state support or external capital. Large-scale private and public capital may be attracted to such localities in order to exploit their natural resources and pockets of surplus

damage caused by development projects, agricultural modernization schemes or afforestation is often muted and voiced most
strongly from outside the region. We might term such areas the
clientelist countryside, where local politics is dominated by a
concern for the welfare of the community (e.g. in job creation,
social housing etc.) and an ambivalence about external state
finance but a strong dependence on it. It is reluctantly accepted,
risks included, because there seem to be few development alternatives if progressive economic decline and social marginalization
are to be avoided.

Conclusion

The four types of countryside discussed above indicate the range
of conditions that may result from the operation of similar
economic, social and political forces. They may offer quite different local responses to environmental conflicts, inward capital
investment and the balance struck between employment generation and the protection of the countryside. They illustrate the
extent to which, in the "differentiated countryside", outcomes are
dependent upon the local and regional contexts, and the particular situated practices of actor-networks operating within them.
For instance, the landowner's ability to initiate development may
be much more severely curtailed in the "preservationist" countryside than in the "paternalistic" or "clientelistic" countryside. This
much is clear from a comparison of our "ideal types". But "ideal
types" represent no more than a static notion of local conditions
and how these conditions may mediate the broader tendencies
identified previously.

To understand how and why landowners and other social
actors behave as they do, we need a much more dynamic
methodology than that provided by the ideal type. In Chapters
6 and 7 we adopted the maxim "follow the actors" – as they
make their decisions, define their interests and pursue them in
concrete situations. This methodology seems to us essential if we
are to understand rural change in an increasingly differentiated
countryside. We cannot simply "read off" action from some pre-
given rural social structure or rural economy. In response, we
have designed a conceptual approach and methodology that
demands that we attempt, directly, to "capture" the nature of
rural change. "Action-in-context" is an attempt to take seriously
the practices of the key actors while recognizing that they are not

free to do as they wish. We must "follow" actors as they enter into and construct networks of social relations. These networks give rise to the processes of economic, social and political change that form the rural and provide the context for the next round of restructuring.

In spite, therefore, of the increasing global tendencies evident in economic organization, environmental concern, information flow, and political structures, nation states are confronted by growing local and regional disparities that demand, in turn, that centralized systems of regulation give way to local and regional systems. It is therefore hardly surprising that no coherence can be identified in the post-productivist phase of rural development. Local unevenness is its quintessential and necessary feature.

We must examine how this unevenness is produced. We need to build our knowledge of current circumstance from the examination of "real" processes in real places. At one level we can identify global tendencies but these only ever "exist" in particular places at particular times. The processes of commoditization and representation, placed in the context of a shift from production to consumption and a given distribution of property rights, are reshaping and redefining rural localities. So rurality, too, is uneven. It is differentiated across space and time. In order to capture key features of these processes we have further argued for analysis of the land development process as a "window" through which we might most usefully assess rural change. We may follow the actors (persons, institutions) through the process. This allows us to build institutional maps around the various development processes as we trace how categories and linkages are imposed by some actors upon others. The outcome of this process, and the context it provides for others, will, for all the reasons outlined above, be an increasingly differentiated countryside.

Accounting for this differentiation demands the application of the appropriate theoretical tools and comparative research methodologies. In the preceding chapters we have carefully considered what these should be and have indicated how they might be utilized to understand contemporary patterns of change. The task now is to use our theories and methods in order to engage fruitfully with the messy complexities of the new rural spaces.

BIBLIOGRAPHY

Entries set in **bold type** are edited volumes cited more than once.

Advisory Council for Agriculture and Horticulture in England and Wales. 1978. *Agriculture and the countryside* (The Strutt Report). London: MAFF.

Aglietta, M. 1979. *A theory of capitalist regulation*. London: New Left Books.

Allanson, P. 1992. *The MacSharry Plan: modulation in the cereals sector*. ESRC Countryside Change Initiative. Working Paper 27. Newcastle upon Tyne: The University.

Ambrose, P. 1977. *The determinants of land use change*. Milton Keynes, England: Open University Press.

Ambrose, P. 1986. *Whatever happened to planning?* Andover, England: Methuen.

Ambrose, P. & R. Colenutt 1973. *The property machine*. Harmondsworth, England: Penguin.

Anderson, M. 1981. Planning policies and development control in the Sussex Downs AONB. *Town Planning Review* **52**, 5–25.

Archer, M. 1982. Morphogenesis versus structuration: on combining structure and action. *British Journal of Sociology* **33**, 455–83.

Association of County Councils 1990. *The case for the counties*. December.

Astor, Viscount & B. S. Rowntree 1938. *British agriculture: the principles of future policy*. London: Longmans, Green.

Ball, M. 1983. *Housing policy and economic power: the political economy of owner occupation*. London: Methuen.

Ball, M. 1985. Coming to terms with owner occupation. *Capital and Class* **24**, 15–44.

Ball, M. 1986a. Housing analysis: time for theoretical refocus. *Housing Studies* **1**, 147–65.

Ball, M. 1986b. The built environment and the urban question. *Environment and Planning D* **4**, 447–64.

Barlow, J. 1986. Landowners, property ownership and the rural locality. *International Journal of Urban and Regional Research* **10**, 309–29.

Barlow, J. 1988. The politics of land into the 1990s: landowners, developers and farmers. *Policy and Politics* 16, 111–23.

Barlow J. & A. King 1992. The state, the market, the competitive strategy, the housebuilding industry in the United Kingdom, France and Sweden. *Environment and Planning A* 24, 317–462.

Barlow, Sir M. 1940. Royal Commission on the Distribution of the Industrial Population. *Report*. London: HMSO.

Barlow, T. & M. Savage 1988. The politics of growth: cleavage and conflict in a Tory heartland. *Capital and Class*, 156–82.

Barnes, I. 1991. Post-Fordist people: cultural meanings of new techno-economic systems. *Futures*, November.

Barrett, S. & P. Healey (eds) 1985. *Land policy: problems and alternatives*. Aldershot, England: Gower.

Barrett, S., M. Stewart, J. Underwood 1978. *The land market and the development process: a review of research and policy*. Bristol: School of Advanced Urban Studies.

Bateman, J. 1883. *The great landowners of Great Britain and Ireland*, 1971 reprint. Leicester, England: Leicester University Press.

Bell, C. & H. Newby 1971. *Community studies*. London: Allen & Unwin.

Best, R. H. 1981. *Land-use and living space*. London, Methuen.

Blowers, A. 1980. *The limits to power: the politics of local planning policy*. Oxford: Pergamon.

Blowers, A. 1987. Transition or transformation? Environmental policy under Thatcher. *Public Administration* 65, 277–90.

Blunden, J. & N. Curry (eds) 1988. *A future for our Countryside*. Oxford: Basil Blackwell.

Boccara, P. 1985. *Intervenir dans les gestions avec de nouveaux critères*. Paris: Messidor/Editions Sociales.

Boddy, M. 1981. The property sector in late capitalism: the case of Britain. In *Urbanization and urban planning in capitalist society*, M. Dear & A. Scott (eds),. Andover, England: Methuen.

Body, R. 1982. *Agriculture: the triumph and the shame*. London: Temple Smith.

Bohman, J. 1991. *New philosophy of social science: problems of indeterminacy*. Cambridge: Polity Press

Booth, A. 1990. *British economic policy, 1931–49: was there a Keynesian revolution?* Hemel Hempstead: Harvester Wheatsheaf.

Booth, C. 1899. *Life and labour of the people* [2 vols]. London: Williams & Norgate.

Boucher, S., A. Flynn, P. Lowe 1991. The politics of rural enterprise: a British case study. See Whatmore et al. (1991), 120–40.

Bowers, J. K. & P. Cheshire 1983. *Agriculture, the countryside and land-use: an economic critique*. London: Methuen.

Bowlby, S., J. Foord, L. McDowell, 1986. The place of gender in locality studies. *Area* 18, 327–31.

Bradley, T. & P. Lowe (eds) 1984. *Locality and rurality: economy and society in rural regions*. Norwich, England: GeoBooks.

Brindley, T., Y. Rydin, G. Stoker 1989. *Remaking planning: the politics of urban change in the Thatcher years*. London: Unwin Hyman.

Bromley, D. 1991. *Environment and economy: property rights and public policy*. Oxford: Basil Blackwell.

Brotherton, I. 1982. Development pressures and control in the National Parks. *Town Planning Review* **53**, 439–59.

Bruton, M. & D. Nicholson 1985. *Local planning in practice*. London: UCL Press (originally published by Hutchinson).

Bryant, C., L. Russwurm, A. McLellan 1982. *The city's countryside: land and its management in the rural/urban fringe*. Harlow, England: Longman.

Bryden, J. & G. Houston 1978. *Agrarian change in the Scottish Highlands*. London: Martin Robinson.

Burton, T. 1991. Planning and compensation bill. ECOS **12**, 70–51.

Buttel, F., D. Larson, G. Gillespie 1990. *The sociology of agriculture*. New York: Greenwood Press.

Byman, W. J. 1990. New technologies in the agro-food system and US–EC trade relations. See Lowe et al. (1990), 147–67.

Cadman, D. & L. Austin-Crowe 1978. *Property development*. London: Spon.

Callon, M. 1986. Some elements of a sociology of translation: domestication of the scallops and the fishermen of St Brieuc Bay. See Law (ed.), 196–233. London: Routledge & Kegan Paul.

Callon, M. 1991. Techno-economic networks and irreversibility. See Law (1991), 132–64.

Callon, M., J. Law, A. Rip 1985. *Texts and their powers: mapping the dynamics of science and technology*. London: Macmillan.

Callon, M., J. Law & A. Rip (eds) 1986. *Mapping the dynamics of science and technology*. Basingstoke, England: Macmillan.

Carruthers, S. P. (ed.) 1986. *Alternative enterprises for agriculture in the UK* (CAS Report II). Reading, England: University of Reading.

Carter, I. 1974. The Highlands of Scotland as an underdeveloped region. In *Sociology and development*. E. de Kadt & G. Williams (eds), 279–311. London: Tavistock.

Carter, I. 1979. *Farm life in northeast Scotland, 1870–1914: the poor man's country*. Edinburgh: John Donald.

Cawson, A. 1986. Meso-corporatism and industrial policy. *ESRC Corporatism and Accountability Newsletter* **4**, 1–3.

Chambers, I. 1990. *Border dialogues: journeys in postmodernity*. London: Routledge.

Champion, A. G. (ed.) 1989. *Counterurbanization: the changing pace and nature of population deconcentration*. London: Edward Arnold.

Champion, A. G., A. J. Fielding, D. E. Kettle 1989. Counterurbanization in Europe. *Geographical Journal* **155**, 52–80.

Champion, A. G. & A. K. Townsend 1990. *Contemporary Britain: a geographical perspective*. London: Edward Arnold.

Cherry, G. E. 1975. *Peacetime history: environmental planning – national parks and recreation in the countryside*. London: HMSO.

Cheshire, P. & D. Hay 1990. *Urban problems in western Europe*. London: Unwin Hyman.

Clark, G. 1990. Location, management strategy and workers' pensions. *Environment and Planning* 22, 150–76.

Clark, J & P. Lowe 1992. Cleaning up agriculture: Environment, technology and social science. *Sociologia Ruralis* 32, 11–29.

Clegg, S. 1989. *Frameworks of power*. London: Sage.

Clemenson, H. A. 1982. *English country houses and landed estates*. London: Croom Helm.

Cloke, P. (ed.) 1989. *Rural land-use planning in developing nations*. London: Unwin Hyman.

Cloke, P. 1990. Community development and political leadership in rural Britain. *Sociologia Ruralis* 25, 305–22.

Cloke, P. & J. Little 1990. *The rural state: limits to planning in rural society*. Oxford: Clarendon Press.

Cloke, P. & N. Thrift 1990. Class and change in rural Britain. See Marsden et al. (1990).

Clyde Mitchell, J. 1983. Case and situation analysis. *Sociological Review* 31, 187–211.

Cohen, A. 1988. *Belonging*. Manchester: Manchester University Press.

Collins, H. & S. Yearley 1992. Epistemological chicken. In *Science as practice and culture*, A. Pickering (ed.), 301–26. Chicago: University of Chicago Press.

Commission of the European Communities 1988. *The future of rural society*. Commission Communication of 29 July 1988, COM (88) 371.

Connell, P. 1974. *The end of tradition*. London: Routledge & Kegan Paul.

Conservative Party 1989. Full speech by Rt Hon. Christopher Patten MP (Bath), Secretary of State for the Environment, made at the 106th Conservative Party Conference at the Winter Gardens, Blackpool, on Wednesday, 11 October 1989, Conservative Party News Service.

Cooke, P. 1989a. Cultural cosmopolitanism: urban and regional studies into the 1990s. *Geoforum* 20, 241–52.

Cooke, P. 1989b. Locality, economic restructuring and world development. See Cooke (1989c).

Cooke, P. (ed.) 1989c. *Localities*. London: Unwin Hyman.

Country Landowners Association Press Release, 28 February 1990.

Country Landowners Association Press Release, 30 January 1992.

Countryside Review Panel 1987. *New opportunities for the countryside*. Cheltenham, England: Countryside Commission 1987.

Cox, G., A. Flynn, P. Lowe, M. Winter 1988. *Alternative uses of agricultural land in England and Wales*. Berlin: Science Centre.

Cox, G. & P. Lowe 1983. A battle not the war: the politics of the Wildlife and Countryside Act. *Countryside planning yearbook*, Andrew

Gilg (ed.), **4**, 48–76. Norwich: Geo Books.

Cox, G., P. Lowe, M. Winter 1985. Land-use conflict after the Wildlife and Countryside Act 1981: The rôle of the Farming and Wildlife Advisory Group. *Journal of Rural Studies* 1, 173–84.

Cox, G., P. Lowe, M. Winter (eds) 1986. Agriculture: people and policies. London: Allen & Unwin.

Cox, G., P. Lowe, M. Winter 1986a. From state direction to self regulation: The historical development of corporatism in British agriculture. *Policy and Politics* **14**, 475–90.

Cox, G., P. Lowe, M. Winter 1986b. Agriculture and conservation in Britain: a policy community under siege. See Cox et al. (1986), 181–215.

Cox, G., P. Lowe, M. Winter 1987. Farmers and the state: a crisis for corporatism. *Political Quarterly* **58**, 73–81.

Cox, G., P. Lowe, M. Winter 1988. Private rights and public responsibilities: the prospects for agricultural and environmental controls. *Journal of Rural Studies* **4**, 323–37.

Cox, G., P. Lowe, M. Winter 1989. The farm crisis in Britain. See Goodman & Redclift (1989), 113–34.

Cox, G., P. Lowe, M. Winter 1990. *The voluntary principle in conservation: the Farming and Wildlife Advisory Group.* Chichester, England: Packard.

Cox, K. & A. Mair 1991. From localized social structures to localities as agents. *Environment and Planning A* **23**, 197–213.

Council for the Preservation of Rural England 1989. *Concrete objections.* London: CPRE.

Council for the Preservation of Rural England, National Housing and Town Planning Council, and Association of District Councils 1990. *Planning control over farmland: reforming permitted development rights in the countryside.* London: ADC.

Council for the Preservation of Rural England 1992. Press Release 22, January.

Cullingworth, J. B. 1974. *Town and country planning in Britain*, 5th edn. London: Allen & Unwin.

Dalton, H. 1962. *High tide and after: memoirs, 1945–1960.* London: Muller.

Day, G. & J. Murdoch 1993. Locality and community: coming to terms with place. *Sociological Review* (in press).

Day, G., G. Rees, J. Murdoch 1989. Social change, rural localities and the state: the restructuring of rural Wales. *Journal of Rural Studies* **5**, 227–44.

Dicken, P. 1992. *Global shift: The internationalization of economic activity*, 2nd edn. London: Paul Chapman.

District Planning Officers Society 1989. *Report on the adoption of district-wide development plans.* District Planning Officers Society.

Dobson, A. P. 1986. *US wartime aid to Britain, 1940–1946*. Beckenham: Croom Helm.

DOE (Department of the Environment) 1980a. *Development control policy and practice*. Circular 22/80.

DOE 1980b. *Land for private housebuilding*. Circular 9/80.

DOE 1984a. *Green belts*. Circular 14/84.

DOE 1984b. *Industrial development*. Circular 16/84.

DOE 1984c. *Land for housing*. Circular 15/84.

DOE 1986. *Development of small business*. London: HMSO

DOE 1987. *Development involving agricultural land*. Draft circular.

DOE 1988a. *Rural enterprise and development*. Planning Policy Guidance Note 7. London: HMSO.

DOE 1988b. *Green belts*. PPG 2. London: HMSO.

DOE 1989a. *Efficient planning*. A Consultation Paper. July.

DOE 1989b. *Permitted use rights in the countryside*. A Consultation Paper. May.

DOE 1990a. News Release, County Structure Plans to Be Retained Announces Chris Patten, 24 September.

DOE 1990b. *Planning control over agricultural and forestry buildings*. A Consultation Paper. October.

DOE 1992a. *Housing*. PPG 3 (revised). London: HMSO.

DOE 1992b. *Development plans and regional planning guidance*. PPG 12. London: HMSO.

DOE 1992c. *General policy and principles*. PPG 1. London: HMSO.

DOE 1992d. *The countryside and the rural economy*. Revised PPG 7. London: HMSO.

Drewett, R. 1973. The developer's decision process. In *The containment of urban England, vol. 2: the planning system objectives, operations, impacts*, P. Hall, H. R. Gracey, R. Drewett & R. Thomas (eds). London: Allen & Unwin.

Drummond, I. M. 1972. *British economic policy and the empire 1919–1939*. London: Allen & Unwin.

Duncan, S. & M. Savage 1991. Commentary. *Environment and Planning A* **23**, 155–64.

Elson, M. 1986. *Green belts: conflict mediation in the urban fringe*. London: Heinemann.

Elster, J. 1985. *Making sense of Marx*. Cambridge: Polity Press.

European Commission, 1988. *The future of rural society*. Com (88) 601. Brussels: Commission of the European Communities.

European Commission 1991. The development and future of the CAP. Reflections Paper to the Commission, EC Documentation Commission.

Exterretta, M. 1992. Transformation of the labour system and work processes in a rapidly modernising agriculture: the evolving case of Spain. See Marsden et al. (1992).

Falk, W. & T. Lyson 1988. *High tech, low tech, no tech: recent industrial and occupational change in the system*. New York: State University of New York Press.

Featherstone, M. 1991. *Consumer culture and postmodernism*. London: Sage.

Flynn, A. 1986. Agricultural policy and party politics in post-war Britain. See Cox et al. (1986), 216–36.

Flynn, A. 1989. Rural working class interests in party policy making in post-war England. Unpublished PhD thesis, University of London.

Flynn, A. & P. Lowe 1992. The greening of the Tories: the Conservative Party and the environment. In *Green politics two*, W. Rüdig (ed.), 9–36. Edinburgh: Edinburgh University Press.

Flynn, A., P. Lowe & G. Cox 1990. *The rural land development process*. ESRC Working Paper 6. Newcastle upon Tyne: the University.

Fothergill, S. & G. Gudgin 1982. *Unequal growth: urban and regional employment change in the UK*. London: Heinemann.

Friedland, W. & E. Pugliese 1989. Class formation and decomposition in modern capitalist agriculture: comparative perspectives. *Sociologia Ruralis* **14**, 86–92.

Friedmann, H. 1988. Family wheat farms and Third World diets: a paradoxical relationship between waged and unwaged labour. In *Work without wages: comparative studies of housework and petty commodity production*, J. L. Collins & M. E. Giminez (eds) (forthcoming). Binghamton, NY: State University of New York Press.

Friedmann, H. & P. McMichael 1989. Agriculture and the state system: the rise and decline of national agricultures, 1870 to the present. *Sociologia Ruralis* **29**, 93–117.

Fuguitt, G. V. 1985. The non-metropolitan population turnaround. *Annual Review of Sociology* **11**, 259–80.

Gamble, A. 1988. *The free economy and the strong state*. London: Macmillan.

Gertler, M. 1988. The limits of flexibility: comments on the post-Fordist visions of production and its geography. *Institute of British Geographers, Transactions* **13**, 419–32.

Gertler, M. & E. Schoenberger 1992. Industrial restructuring and continental trade blocs: the European Community and North America. *Environment and Planning A* **24**, 2–11.

Giddens, A. 1984. *The constitution of society*. London: Polity Press.

Giddens, A. 1990. *The consequence of modernity*. London: Polity Press.

Gilbert J. & C. Howe 1991. Beyond state versus society: theories of the state and New Deal agricultural policies. *American Sociological Review* **56**, 204–20.

Glyn, A. 1989. The crash and real capital accumulation. *Capital and Class* **3**, 21–4.

Goffman, E. 1974. *Frame analysis: an essay on the organization of experience*.

New York: Harper & Row.

Goodchild, R. & R. J. C. Munton 1985. *Development and the landowner: an analysis of the British experience*. London: Allen & Unwin.

Goodman, D. & M. Redclift (eds) 1989. *The international farm crisis*. **London: Macmillan.**

Goodman, D. & M. Redclift 1991. *Refashioning nature: food, ecology and culture*. London: Routledge.

Goodman, D., B. Sorj, J. Wilkinson 1987. *From farming to biotechnology: a theory of agro-industrial development*. Oxford: Basil Blackwell.

Gordon, D. M. 1980. Stages of accumulation and long economic cycles. In *Processes of the world-system*, T. Hopkins & I. Wallerstein (eds), 9–45. Beverly Hills: Sage.

Gore, T. & D. Nicholson 1985. Alternative frameworks for the analysis of public sector land ownership and development. See Barrett & Healey (1985),. Aldershot: Gower.

Gore, T. & D. Nicholson 1991. Models of the land-development process: a critical review. *Environment and Planning A* 23, 705–30.

Grant, M. 1988. *Urban planning law: first supplement*. London: Sweet & Maxwell.

Grant, M. 1990. *Urban planning law: second supplement*. London: Sweet & Maxwell.

Greenwell Working Party Report 1989. *Enterprise in the rural environment*. London: Country Landowners Association.

Gregson, N. 1987. *Locality research: a case of conceptual duplication*. Discussion Paper 86. Newcastle upon Tyne: Centre for Urban and Regional Studies, the University.

Grigg, D. 1987. Farm size in England and Wales from early Victorian times to the present. *Agricultural History Review* 35, 179–89.

Grigg, D. 1989. *English agriculture: an historical perspective*. Oxford: Basil Blackwell.

Habermas, J. 1976. *Legitimation crisis*. London: Heinemann.

Hall, A. D. 1941. *Agriculture after the war*. London: John Murray.

Hall, P., R. Thomas, H. Gracey & R. Drewett 1973. *The containment of urban England*. London: Allen & Unwin.

Hamnett, C. 1987. The Church's many mansions: the changing structure of the Church Commissioners' land and property holdings, 1948–1977. *Institute of British Geographers, Transactions* 12, 465–81.

Harris, K. 1982. *Attlee*. London: Weidenfield & Nicolson.

Harrison, M. L. 1987. Property rights, philosophies, and the justification of planning control. In *Planning control: philosophies, prospects and practices*, M. L. Harrison & R. Mordey (eds), 32–58. London: Croom Helm.

Harvey, D. 1978. The urban process under capitalism: a framework for analysis. *International Journal of Urban and Regional Research* 2, 101–31.

Harvey, D. 1982. *The limits to capital*. Oxford: Basil Blackwell.

Harvey, D. 1985. *The urbanization of capital*. Oxford: Basil Blackwell.

Harvey, D. 1989. *The condition of post-modernity*. Oxford: Basil Blackwell.

Healey, P. 1991. Urban regeneration and the development industry. *Regional Studies* 25, 97–110.

Healey. P. & S. Barrett 1990. Structure and agency in land and property development processes: some ideas for research. *Urban Studies* 27, 89–104.

Healey, P., P. MacNamara, M. Elson & A. Doak (eds) 1988. *Land use planning and the mediation of urban change: the British planning system in practice*. Cambridge: Cambridge University Press.

Held, D. 1991. Democracy, the nation state and the globalisation. *Economy and Society* 20, 138–73.

Hill, B. & R. Gasson 1985. Farm tenure and farming practice. *Journal of Agricultural Economics* 35, 187–99.

Hindess, B. 1986a. Actors and social relations. In *Sociological theory in transition*, M. Wardell & S. Turner (eds). London: Allen & Unwin.

Hindess, B. 1986b. Interests in political analysis. In *Power, action and belief: a new sociology of knowledge?*, J. Law (ed.), 112–31. London: Routledge & Kegan Paul.

Hindess, B. 1988. *Choice, rationality and social theory*. London: Unwin Hyman.

Hirsch, F. 1976. *The social limits to growth*. Cambridge, Mass.: Harvard University Press.

Hirst, P. & J. Zeitlin 1991. Flexible specialisation versus post-Fordism: theory, evidence and policy implications. *Economy and Society* 20, 1–56.

Hobsbawm, E. 1975. *The age of capital, 1848–1875*. London: Weidenfeld & Nicolson.

Hodge, I.D. & S. Monk, 1987. Manufacturing employment change within rural areas. *Journal of Rural Studies* 3, 65–9.

Hoffman, G.W. 1932. *Future trading upon organised commodity markets to the United States*. Philadelphia: University of Pennsylvania Press.

Home, R. 1987. *Planning use classes*. Oxford: Blackwell Scientific.

Hooper, A., P. Pinch & S. Rogers 1988. Housing land availability: circular advice, circular arguments and circular methods. *Journal of Planning and Environmental Law*, 225–39.

House of Lords 1990. *The future of rural society*. Report from the Select Committee on the European Communities. London: HMSO.

Jacques, M. & S. Hall (eds) 1989. *New times*. London: Lawrence & Wishart.

Jenkins, T. N. 1990. *Future harvests: the economics of farming and the environment - proposals for action*. London: CPRE/WWF.

Jessop, B. 1990. Regulation theories in retrospect and prospect. *Economy and Society* 19, 153–216.

Jessop, B., K. Bonnett, S. Bromley, T. Ling 1989. *Thatcherism: a tale of two nations*. Cambridge: Polity Press.

Johnson, C. 1990. Farmland as a business asset. *Journal of Agricultural Economics* **41**, 135–48.

Johnston, R. J. 1991. *A question of place: exploring the practice of human geography*. Oxford: Basil Blackwell.

Jones Lang Wootton 1983. *The agricultural land market in Britain: an in-depth study of its mechanism for owners, investors and advisors*. London: Jones Lang Wootton Research.

JURUE 1982. *The planning problems of small firms in rural areas*. Birmingham: Joint Unit for Research in the Urban Environment, University of Aston.

Keeble, D. E., P. L. Owens, C. Thompson, 1983. The urban–rural manufacturing shift in the European Community. *Urban Studies* **20**, 405–18.

Kenney, M., L. Lobao, J. Curry, W. R. Goe 1989. Mid-western agriculture in US Fordism, from the new deal to economic restructuring. *Sociologia Ruralis* **29**, 131–49.

Kneale, J., P. Lowe, T. K. Marsden 1992. The conversion of agricultural buildings: an analysis of variable pressures and regulations towards the post-productivist countryside. ESRC Countryside Change Initiative, Working Paper 29. Newcastle upon Tyne: the University.

Knorr-Cetina, K. 1988. The micro-social order: towards a reconception. In *Action and structure*, N. Fielding (ed.), 21–53. London: Sage.

Kolko, G. 1968. *The politics of war: the world and the United States foreign policy, 1943–1945*. New York: Vintage Books and Random House.

Kolko, J. & G. Kolko 1972. *The limits of power: the world and the United States foreign policy, 1945*. New York: Harper & Row.

Land Agents' Society 1916. *Facts about land: a reply to 'The land', the report of the Unofficial Land Enquiry Committee*. London: Land Agents' Society.

Land Enquiry Committee 1913. *The land: the report of the Land Enquiry Committee*. London: Hodder & Stoughton.

Lash, S. & J. Urry 1987. *The end of organised capitalism*. Cambridge: Polity Press.

Latour, B. 1987. *Science in action*. Milton Keynes, England: Open University Press.

Latour, B. 1991. Technology is society made durable. See Law (1991), 103–31.

Law, J. 1986a. Editor's introduction: power/knowledge and the dissolution of the sociology of knowledge. See Law (1986b), 1–19.

Law, J. (ed.) 1986b. *Power, action and belief: a new sociology of knowledge?* **London: Routledge & Kegan Paul.**

Law, J. 1991. *A sociology of monsters: essays on power, technology and domination*. **London: Routledge.**

Lawrence, G. 1990. Agricultural restructuring and rural social change in Australia. See Marsden et al. (1990), 101–28.

Le Heron, R. 1991. New Zealand agriculture and changes in the agriculture–finance relation during the 1980s. *Environment and Planning A* **23**, 1653–70.

Lee, J.M. 1963. *Social leaders and public persons: a study of county government in Cheshire since 1888*. Oxford: Oxford University Press.

Leyshon, A. & N. Thrift 1988. The gambling propensity: banks, developing country debt exposures and the new international financial system. *Geoforum* **19**, 55–69.

Leyshon, A. & N. Thrift 1992. Liberalisation and consolidation: The single European market and the remaking of European finance capital. *Environment and Planning A* **24**, 49–83.

Lichfield, N. 1956. *The economics of planned development*. London: Estates Gazette.

Lipietz, A. 1985. The world crisis: the globalization of the general crisis of Fordism. *IDS Bulletin* **16**, 6–20.

Lipietz, A. 1987. *Mirages and miracles: the crises of global Fordism*. London: Verso.

Lipietz, A. 1990. The debt problem, European integration and the new phase of world crisis. *New Left Review*, 37–51.

Long, N. 1989. Theoretical reflections on actor, structure and interface. In *Encounters at the interface: a perspective on social discontinuities in rural development*, N. Long (ed.) Wageningse Sociologische Studies 27, 221–44.

Long, N. 1990. From paradigm lost to paradigm regained? The case for an actor-oriented sociology of development. *European Review of Latin American and Caribbean Studies* **49**, 3–24.

Long, N. & J. D. van der Ploeg 1988. New challenges in the sociology of rural development: a rejoinder to Vandergeest. *Sociologia Ruralis* **18**, 30–42.

Long, N. & J. D. van der Ploeg 1989. Demythologising planned intervention: an actor perspective. *Sociologia Ruralis* **19**, 227–49.

Long, N., J. van der Ploeg, C. Curtin, L. Box (eds) 1986. *The commoditisation debate: labour process, strategy and social network*. Working Paper No. 17, University of Wageningen.

Lowe, P. & M. Bodiguel (eds) 1990. *Rural studies in Britain and France*. London: Pinter (Belhaven).

Lowe, P., G. Cox, T. O'Riordan, M. MacEwen, M. Winter 1986. *Countryside conflicts: the politics of farming, forestry and conservation*. Aldershot: Gower

Lowe, P., & J. Goyder 1983. *Environmental groups in politics*. London: Allen & Unwin.

Lowe, P., T. K. Marsden, S. Whatmore (eds) 1990. *Technological change and the rural environment*. **London: David Fulton.**

Lowe. P. & M. Winter 1987. Alternative perspectives on the alternative land use debate. In *Farm extensification*, N. R. Jenkins & M. Bell (eds).

Grange-over-Sands, England: Institute for Terrestrial Ecology.

McAuslan, P. 1980. *The ideology of planning law.* Oxford: Pergamon.

MacEwen, A. & M. MacEwen 1982. *National parks: conservation or cosmetics.* London: Allen & Unwin.

Mckay, D. & A. Cox 1979. *The politics of urban change.* London: Croom Helm.

McLoughlin, B. 1989. *New uses for agricultural land.* London: NFU (typescript).

McMichael, P. 1984. *Settlers and the agrarian question: foundations of capitalism in colonial Australia.* New York: Cambridge University Press.

McMichael. P. 1985. Britain's hegemony in the nineteenth-century world-economy. In *States versus markets in the world-system,* P. Evans, D. Rueschmeyer, E. H. Stephens (eds), 117–50. Beverly Hills, Calif.: Sage.

McMichael, P. (ed.) 1992. *Agro-food system restructuring in the late twentieth century: comparative and global perspectives.* Ithaca, NY: Cornell University Press.

McMichael, P. & D. Myhre 1991. Global regulation versus the nation state. Agro-food systems and the new politics of capital. *Capital and Class* **43**, 83–107.

Major, J. 1992. Speech to the Annual Conference of the National Farmers' Union. Oxford.

Marsden, T. K. 1986. Property–state relations in the 1980s: an examination of landlord–tenant legislation in British agriculture. See Cox et al. (1986c), 126–45.

Marsden, T. K., P. Lowe, S. Whatmore (eds) 1990. *Rural restructuring: global processes and their responses.* London: David Fulton.

Marsden, T. K., P. Lowe, S. Whatmore (eds) 1992. *Labour and locality: uneven development in the labour process.* London: David Fulton.

Marsden, T. K., R. J. C. Munton, S. Whatmore, J. K. Little 1986. Towards a political economy of agriculture: a British perspective. *International Journal of Urban and Regional Research* **11**, 498–521.

Marsden, T. K. & J. Murdoch, K. Sullivan, V. Lingham 1992. Planning for social limits to growth. *The Planner* **78**, 4, 6–7.

Marsden, T. K., J. Murdoch, S. Williams 1991. *From farmers to developers: contested transitions in the development of small industrial units on farms in a prosperous rural region.* ESRC Countryside Change Initiative, Working Paper 22. Newcastle upon Tyne: the University.

Marsden, T. K., N. Ward & R. J. C. Munton 1992. Farm businesses in upland and lowland Britain: incorporating social trajectories into uneven agrarian development. *Sociologia Ruralis.*

Marsden, T. K. & S. Whatmore 1992. Finance capital and food system restructuring: global dynamics and their national incorporation. In *Food systems and agrarian change in the late twentieh century,* P. McMichael (ed.). Ithaca, NY: Cornell University Press.

Marsden, T. K. & S. Williams 1992. Integrated panel report for the Buckinghamshire study area. *Rural change in Europe: research programme on farm structures and pluriactivity.* Nethybridge, Inverness, Scotland: The Arkleton Trust.

Martin, R. 1989. The reorganisation of regional theory: alternative perspectives on the changing capitalist space economy. *Geoforum* 20, 187–201.

Massey, D. 1978. Regionalism: some current issues. *Capital and Class* 6, 106–25.

Massey, D. 1984. *Spatial divisions of labour: social structures and the geography of production.* London: Macmillan.

Massey, D. 1989. Reflections on the debate: thoughts on feminism, Marxism and theory. *Environment and Planning A* 21, 692–7.

Massey, D. 1991. A global sense of place. *Marxism Today,* June, 24–9.

Massey, D. & J. Allen 1988. *The economy in question.* London: Sage.

Massey, D. & A. Catalano 1977. *Capital and land: land ownership by capital in Great Britain.* London: Edward Arnold.

Merrett, S. 1979. *State housing in Britain.* London: Routledge & Kegan Paul.

Milward, A. S. 1984. *The reconstruction of Western Europe, 1945–51.* London: Methuen.

Mingay, G. E. (ed.) 1981. *The Victorian countryside.* London: Routledge & Kegan Paul.

Mingay, G. E. 1990. *A social history of the English countryside.* London: Routledge.

Ministry of Agriculture and Food 1941. *The National Farm Survey of 1941.* London: HMSO.

Ministry of Agriculture, Fisheries and Food 1979. *Farming and the nation.* Cmnd 7458. London: HMSO.

Ministry of Agriculture, Fisheries and Food 1987. *Food from our own Resources.* Cmnd 6020. London: HMSO.

Ministry of Agriculture, Fisheries and Food 1988. *Farm rents in England and Wales: results of the 1987 annual rent inquiry.* London: MAFF.

Mormont, M. The emergence of rural struggles and their ideological effects. *International journal of urban and regional research* 7, 559–78.

Mormont, M. 1990. Who is rural? Or, how to be rural. Towards a sociology of the rural. See Marsden et al. (1990), 21–45.

Mouzelis, N. 1991. *Back to sociological theory: the construction of social orders.* London: Macmillan.

Munton, R. J. C. 1983. *London's green belt: containment in practice.* London: Allen & Unwin.

Munton, R. J. C. 1984. The politics of rural landownership: institutional investors and the Northfield enquiry. See Bradley & Lowe (1984), 167–80.

Munton, R. J. C. 1985. Investment in British agriculture by the financial institutions. *Sociologia Ruralis* **25**, 153–73.

Munton, R. J. C., S. J. Whatmore, T. K. Marsden 1988. Reconsidering urban-fringe agriculture: a longitudinal analysis of capital restructuring on farms in the Metropolitan green belt. *Institute of British Geographers, Transactions* **13**, 324–36.

Murdoch, J. 1992. Representing the region: Welsh farmer and the British state. See Marsden et al. (1992), 160–81.

Murdoch, J. & T. K. Marsden 1991. *Reconstituting the rural in an urban region: new villages for old?* ESRC Countryside Change Initiative, Working Paper 26. Newcastle upon Tyne: the University.

Nalson, J. S. 1968. *Mobility of farm families*. Manchester: Manchester University Press.

National Farmers Union 1989/90. Land use policy review – final report.

National Farmers Union 1992. Annual conference.

National Farmers Union & Country Landowners Association, 1977.

Nature Conservancy Council, 1977. *Nature conservation and agriculture*. London: NCC.

Newby, H. 1977. *The deferential worker: a study of farm workers in East Anglia*. London: Allen Lane.

Newby, H. 1985. *Green and pleasant land?: social change in rural England*, 2nd edn. London: Hutchinson.

Newby, H. 1986. Locality and rurality: the restructuring of rural social relations. *Regional Studies* **20**, 209–16.

Newby, H. 1987. *Country life: a social history of rural England*. London: Weidenfeld & Nicolson.

Newby, H., C. Bell, D. Rose, P. Saunders 1978. *Property, paternalism and power: class and control in rural England*. London: Hutchinson.

Northfield Committee 1979. *Report of the committee of inquiry into the acquisition and occupancy of agricultural land*. Cmnd 7599. London: HMSO.

Norton-Taylor, R. 1982. *Whose land is it anyway?* Wellingborough, England: Turnstone Press.

O'Cinnéide, M. & M. Cuddy (eds) 1992. *Perspectives on rural development in advanced economies*. Galway, Ireland: Centre for Development Studies, University College, Galway.

O'Connor, J. 1984. *Accumulation crisis*. Oxford: Basil Blackwell.

Offe, C. 1984. Crises of crisis management: elements of political crisis theory. In *Contradictions of the welfare state*, J. Keane (ed.) 35–65. London: Hutchinson.

Offer, A. 1981. *Property and politics, 1870–1914: landownership, law, ideology and urban development in England*. Cambridge: Cambridge University Press.

Orwin, C. S. 1949. *A history of English farming*. Edinburgh: Thomas Nelson.

Pahl, R. 1965. *Urbs in rure: the metropolitan fringe of Hertfordshire.* London: Weidenfeld & Nicolson.

Patten C. 1989. *Planning and local choice.* Speech given 4 October 1988, Department of the Environment.

Peck, J. & A. Tickell 1992. *Local modes of social regulation? Regulation theory, Thatcherism and uneven development.* University of Manchester, Department of Geography, Spatial Policy Analysis Working Paper 14.

Perring, F. H. & K. Mellanby (eds) 1977. *Ecological effects of pesticides.* London: Academic Press.

Persson, L. O. 1992. Rural labour markets meeting urbanisation and the arena society: new challenges for policy and planning in rural Scandinavia. See Marsden et al. (1992), 68–95.

Peterson, M. 1990. Paradigmatic shift in agriculture: global effects and the Swedish response. See Marsden et al. (1990).

Pettigrew, P. 1987. A bias for action: industrial development in mid-Wales. In *Rural planning: policy into action*, P. Cloke (ed.). London: Harper & Row.

Pickvance, C. & E. Preteceille (eds) 1991. *State restructing and local power: a comparative perspective.* London: Pinter.

Pile, S. 1990. *The private farmer: transformation and legitimation in advanced capitalist agriculture.* Aldershot, England: Dartmouth.

Pimlott, B. 1985. *Hugh Dalton.* London: Jonathan Cape.

Piore, M. & C. F. Sabel 1984. *The second industrial divide.* New York: Basic Books.

The Planner 1992. Editorial. *Royal Town Planning Institute, Journal.*

van der Ploeg, J. D. 1990. *Labour, markets and agricultural production.* Oxford: Westview Press.

van der Ploeg, J. D. 1992. The reconstitution of locality: technology and labour in modern agriculture. See Marsden et al. (1992),.

Poloscia, R. 1991. Agriculture and diffused manufacturing in the Terza Italia: a Tuscan case study. See Whatmore et al. (1991), 34–58.

Potter, C. 1991. *The diversion of land: conservation in a period of farming contraction.* London: Routledge.

Reade, E. 1987. *British town and country planning system.* Milton Keynes, England: Open University Press.

Rees, G. 1984. Rural regions in national and international economies. See Bradley & Lowe (1984), 27–44.

Rhodes, R. 1986. *The national world of local government.* London: Allen & Unwin.

Rhodes, R. 1988. *Beyond Westminster and Whitehall: the sub-central governments of Britain.* London: Unwin Hyman.

Rooth, T. 1985. Trade agreements and the evolution of British agricultural policy in the 1930s. *Agricultural History Review* **33**, 173–90.

Rowntree, S. 1915. *Poverty report.* Publisher?

Rydin, Y. 1986. *Housing land policy.* Aldershot, England: Gower.

Salamon, S. 1980. Ethnic difference in farm family land transfers. *Rural Sociology* **45**, 290–308.

Saunders, P. 1990. *A nation of home owners*. London: Routledge.

Saunders, P. & C. Harris 1990. Privatisation and the consumer. *Sociology* **24**, 57–75.

Savage, R. 1987. *The future of rural planning and the environmental implications: the rural view* (Council Paper). London: Royal Town Planning Institute.

Savage, M., P. Dicken, M. Fielding 1992. *Property, bureaucracy and culture: middle-class formation in contemporary Britain*. London: Routledge.

Savills–IPD 1989. *Agriculture performance analysis 1988*. London: Savills–IPD.

Scott Committee 1942. *Report on land utilisation in rural areas*. Cmnd 6378. London: HMSO.

Scott, A. MacEwan 1986. Towards a rethinking of petty commodity production. *Social Analysis* **20**, 93–105.

Scrase, A. J. 1988. Agriculture: 1980s industry and 1947 definition. *Journal of Planning and Environmental Law*, July, 447–60.

Secretary of State for the Environment 1985. *Lifting the burden*. London: HMSO.

Secretary of State for the Environment 1989. *The future of development plans*. London: HMSO.

Secretary of State for the Environment 1990. *This common inheritance*. Cmnd 1200. London: HMSO.

Sheail, J. 1981. *Rural conservation in inter-war Britain*. Oxford: Clarendon Press.

Shoard, M. 1980. *The theft of the countryside*. London: Maurice Temple Smith.

Shoard, M. 1987. *This land is our land*. London: Maurice Temple Smith.

Short, J. R. et al. 1986. *Housebuilding, planning and community action*. London: Routledge & Kegan Paul.

Shucksmith, M. 1990. *Housebuilding in Britain's countryside*. London: Routledge.

Silverman, D. 1985. *Qualitative methodology and sociology*. Aldershot: Gower.

Simmie, J. 1974. *Citizens in conflict: the sociology of town planning*. London: Hutchinson.

Smith, C. 1986. Reconstructing the elements of petty commodity production. *Social Analysis* **20**, 41–52.

Smith, M. J. 1989. Changing agendas and policy communities: agricultural issues in the 1930s and 1980s. *Public Administration* **67**, 149–65.

Smith, M. J. 1990. *The politics of agricultural support in Britain*. Aldershot, England: Dartmouth.

Smith, R. 1992. On the margins: uneven development and rural restructuring in the Highlands of Scotland. See Marsden et al. (1992).

Stamp, D. L. 1950. *The land of Britain – its use and misuse*, 2nd edn. London: Longman.

Stanley, O. 1984. *Taxation of farmers and landowners*, 2nd edn. London: Butterworth.

Storper, M. 1986. Capital and industrial location. *Progress in Human Geography* 15, 473–509.

Summers, G., F. Horton & C. Gringeri 1990. Rural labour and market changes in the United States. See Marsden et al. (1990), 129–65.

Sutherland, D. 1988. *The landowners*, 2nd edn. London: Muller.

Symes, D. & J. Appleton 1986. Family goals and kinship strategies in a capitalist farming society. *Sociologia Ruralis* 26, 346–63.

Thompson, F. M. L. 1963. *English landed society in the nineteenth century*. London: Routledge & Kegan Paul.

Thompson, G. 1989. Flexible specialisation, industrial districts, regional economies: strategies for socialists? *Economy in Society* 18, 527–45.

Thrift, N. 1987a. Introduction: the geography of late twentieth century class formation. In *Class and space: the making of urban society*, N. Thrift & P. Williams (eds), 207–53. London: Routledge.

Thrift, N. 1987b. Manufacturing rural geography. *Journal of Rural Studies* 3, 77–81.

Thrift, N. 1989. Images of social change. In *The changing social structure*, C. Hamnett, L. McDowell, P. Sarre (eds),. London: Sage.

Touraine, A. 1974. *Production de la société*. Paris: Seuil.

Tracy, M. 1982. *Agriculture in western Europe: crisis and adaptation since 1880*. London: Jonathan Cape.

Tracy, M. (ed.) 1990. *Rural policy issues*. Nethybridge, Inverness, Scotland: The Arkleton Trust.

Urry, J. 1981. Localities, regions and social class. *International Journal of Urban and Regional Research* 5, 455–74.

Urry, J. 1984. Capitalist restructuring, recomposition and the regions. See Bradley & Lowe (1984), 45–64.

Urry, J. 1990. Conclusion: places and policies. In *Place, policy and politics: do localities matter?* M. Harloe, C. Pickvance, J. Urry (eds), 187–204. London: Unwin Hyman.

Urry, J. 1990. *The tourist gaze*. London: TCS.

Uthwatt, Mr Justice, 1942. Expert committee on compensation and betterment. London: HMSO.

Ward, N. 1990. A preliminary analysis of the UK food chain. *Food Policy* 15, 439–41.

Warde, A. 1985. Spatial change, politics and the division of labour. In *Social relations and spatial structures*, D. Gregory & J. Urry (eds). London: Macmillan.

Warde, A. 1990. Introduction to the sociology of consumption. *Sociology*

24, 1–4.

Watkins, C. & M. Winter 1988. *Superb conversions? Farm diversification – the farm building experience.* London: CPRE.

Westmacott, R. & T. Worthington 1974. *New agricultural landscapes.* Cheltenham, England: Countryside Commission.

Westmacott, R. & T. Worthington 1984. *New agricultural landscapes: a second look.* Cheltenham, England: Countryside Commission.

Whatmore, S. 1988. From women's rôle to gender relations: changing perspectives in the analysis of farm women. *Sociologia Ruralis* **28,** 239–47.

Whatmore, S., P. Lowe, T. K. Marsden (eds) 1991. *Rural enterprise: shifting perspectives on small-scale production.* **London: David Fulton.**

Whatmore, S., R. J. C. Munton, T. K. Marsden 1990. The rural restructuring process: emerging divisions of property rights. *Regional Studies* **24,** 235–45.

Whetham, E. H. 1978. *The agrarian history of England and Wales, volume 8: 1914–1939.* Cambridge: Cambridge University Press.

Whitener, L. A. 1989. The agricultural labour market:a conceptual perspective. In *Research in rural sociology and development,* vol. 4, W. Falk & T. A. Lyson (eds), 55–79. Greenwich, Connecticut: JAI Press

Wiener, M. J. 1981. *English culture and the decline of the industrial spirit 1850–1980.* Cambridge: Cambridge University Press.

Williams, Lord 1965. *Digging for Britain.* London: Hutchinson.

Williams, T. 1935. *Labour's way to use the land.* London: Methuen.

Williams, W. M. 1964. *A West Country village: Ashworthy.* London: Routledge & Kegan Paul.

Williams, R. 1973. *The country and the city.* London: Chatto & Windus.

Winter, M., C. Richardson, C. Short & C. Watkins 1990. *Agricultural land tenure in England and Wales.* London: RICS.

Wormell, P. 1978. *The anatomy of agriculture: a study of Britain's greatest industry.* London: Harper & Row.

INDEX